Maths for Chemistry

A chemist's toolkit of calculations

Paul Monk

School of biology, chemistry & health science
Manchester Metropolitan University

OXFORD

UNIVERSITY PRESS

OXFORD

UNIVERSITY PRESS

Great Clarendon Street, Oxford OX2 6DP

Oxford University Press is a department of the University of Oxford.
It furthers the University's objective of excellence in research, scholarship,
and education by publishing worldwide in

Oxford New York

Auckland Cape Town Dar es Salaam Hong Kong Karachi
Kuala Lumpur Madrid Melbourne Mexico City Nairobi
New Delhi Shanghai Taipei Toronto

With offices in

Argentina Austria Brazil Chile Czech Republic France Greece
Guatemala Hungary Italy Japan Poland Portugal Singapore
South Korea Switzerland Thailand Turkey Ukraine Vietnam

Published in the United States
by Oxford University Press Inc., New York

British Library Cataloguing in Publication Data

Data available

Library of Congress Cataloguing in Publication Data

Data available

Typeset by Newgen Imaging Systems (P) Ltd., Chennai, India
Printed in Great Britain
on acid-free paper by
Ashford Colour Press, Gosport, Hampshire

ISBN 0–19–927741–9   978–0–19–927741–4 (pbk.)

1 3 5 7 9 10 8 6 4 2

Contents

Preface

Many people, when they first start learning advanced and degree-level chemistry, are dismayed when told they must also learn mathematics. They say, 'But I want to be a chemist and *not* a mathematician!' In fact, all chemists use mathematics each day of their working lives. It's one of our most useful tools. And we need to remember that mathematics is only a tool and never a master.

This book is intended for people who want to learn chemistry but either find mathematics very hard, or have never really studied it and therefore need to master its rules. Most people, when they start to learn a musical instrument or a foreign language, initially find great difficulty in remembering the 'mechanics' of their new art. For example, they need to remember how to position their fingers every time they see a b-flat note on a sheet of music; and trainee linguists need to remember how to conjugate a verb ending each time they encounter the verb 'is'. But after a remarkably short period of time, the music student's fingers will automatically reach for the right note, and the student learning a language can conjugate verbs seemingly without effort. Having attained this stage, the musician can concentrate on what the music actually sounds like; and the linguist can concentrate on what the passage to be translated actually means.

Many people reading this book are similar to the trainee musician: we may know *how* to finger some of the notes, but we are not yet proficient enough to take the fingering for granted, and we need to learn more notes and develop our technique. We routinely forget to concentrate on the melody: the point of playing is to make a beautiful sound, not to move the fingers. Similarly, we will be better chemists after we master the simple rules of mathematics, because we will know what the equations actually *mean*. Our first task, then, is to learn the grammar of mathematics.

One of the greater joys in writing a book is the opportunity afforded me to thank the many colleagues who have made it possible. I therefore wish to thank Michael Rycraft (formerly of the Department of Mathematics, MMU) who read the entire manuscript, and made a great many clever comments. Also, Dr Chris Rego and Dr David Johnson, both of whom work in my own department, who each read several chapters. Again, their expertise is evident in the final version of the text. Similarly, I would like to thank the 2004–5 cohort of first-year students at MMU, who acted as 'guinea pigs' in our 'Mathematics for Chemists' unit, in which chapters of this book were issued in the form of weekly handouts. Their comments on the first draft were frank but fair, and eventually appreciative. Any errors and obscurities remaining are of course entirely my own.

I wish to thank Professors Pau Atela and Christophe Gole of the Mathematics Department, Smith College, USA, for permission to reproduce Fig. 6.1.

I would also like to thank Jonathan Crowe, Ruth Hughes and June Cummings, all of Oxford University Press, who have ensured the project's smooth running.

Finally, I thank Dr Paul May, University of Bristol, and Dr Martin Loftus, University of East Anglia, who read almost the entire draft manuscript, and offered numerous valuable comments and suggestions to help refine the text.

Instructions for the tutor

This book is intended to represent a course lasting a single academic year. Accordingly, each chapter will represent about one week's worth of work. One lecture and one tutorial per chapter will probably be an appropriate amount of time. The course was developed to enable the early chapters to be taught without requiring the material from later chapters.

Each chapter is clearly too long for its entire content to be covered in depth during a single lecture. It is suggested therefore that tutors familiarize themselves with the text, and then offer the key points, using a few worked examples for ramification. Students will then be required to read the whole chapter properly in their own time before the respective tutorial.

The contextual approach adopted here is intended to simplify the course, since most students more readily learn material if they already know a 'skeleton' and can add 'flesh' to it, rather than learning from scratch. Occasionally, students will find the chemistry more difficult than the mathematics; unfortunately, the topics deemed difficult will differ from student to student. It is suggested that the tutor regularly reassures the nervous student, saying, 'You only need to understand the examples; it is the mathematics you need to learn.' Only the numbered equations need to be learnt. The text contains about 30 such equations.

The answers to all the self-test questions may be found in the appendices. Most are given as full model answers. The answers to the additional problems are available on the Internet at *www.oxfordtextbooks.co.uk/orc/monk/*, again as fully worked model problems.

Instructions for the student

It is tempting to start a course in mathematics with the words, 'But I want to study *chemistry*!' This course was written with such a mindset in view. The contextual approach here is intended to emphasize how the mathematics is a key tool in the chemist's tool-chest, and not an esoteric 'extra'. Occasionally the chemistry will obscure rather than enlighten. If this is the case for you, jettison the chemistry and concentrate on the maths, or simply go on to the next worked example. You will never be asked in a maths course to remember the chemistry examples. The only equations you will be asked to memorize for this course are those with a number, e.g. the operator Δ is defined in eqn (1.1). There are just over 30 such equations. Most are very easy; in fact, you will probably have met some of them before starting this course.

It is recommended that you read a chapter through from end to end before starting any of the self-test questions. This allows for an overview and, hopefully, a clearer vision of what is involved. Then reread the relevant section before starting the self-test questions. The answers to all these questions may be found in the appendices. Most are given as full model answers. The answers to the additional problems are available on the Internet at *www.oxfordtextbooks.co.uk/orc.monk/*, again as fully worked model problems.

Symbols

Algebraic symbols

a	acceleration; constant; van der Waals constant
a_0	atomic diameter
A	ampere
A	area; Arrhenius pre-exponential factor; Debye–Hückel 'A' factor
Abs	absorbance
b	constant; van der Waals constant
c	intercept on a graph; speed of light; constant; constant of integration; number of components
C	coulomb; Celsius
C	heat capacity
C_p	heat capacity at constant pressure
C_V	heat capacity at constant volume
$[C]$	concentration
d	differential operator
d	interplane distance in a regular crystal; distance
D	diffusion coefficient
Da	Dalton
e	exponential
e^-	electron
E	energy
$\mathcal{E}$	electric field
E_a	activation energy
$E_{O,R}$	electrode potential for the redox couple, $O + ne^- = R$
$E_{O,R}^{\ominus}$	standard electrode potential for the redox couple, $O + ne^- = R$
exp	exponential (i.e. 2.178^x)
f	number of degrees of freedom
$f(x)$	function of x; distribution of x

F	temperature in Fahrenheit
F	Faraday constant
g	gram
g	acceleration due to gravity
G	Gibbs function
h	hour
h	height; the Planck constant; Miller index
$\hbar$	$h \div 2\pi$
H	enthalpy
H_c	enthalpy of combustion
H_f	enthalpy of formation
H_r	enthalpy of reaction
Hz	Hertz
i	item number
I	ionic strength; current
I_t	time-dependent current
j	item number
J	Joule
k	kilo
k	rate constant; bond force constant; abbreviation for (RT/F); Miller index
k_B	Boltzmann constant
k_n	rate constant of the nth step in a reaction series
k_{-n}	rate constant of the reverse of the nth step in a reaction series
K	kelvin
K	equilibrium constant
K_{sp}	equilibrium constant of solubility ('solubility product')
l	Miller index

l	length	T	absolute temperature
ln	logarithm to base e (i.e. natural logarithm)	u	top function of a quotient; first function of a product; substituent in a chain-rule problem; initial velocity
log	logarithm to base 10		
$\log_x$	logarithm to base x	U	internal energy; attractive forces
m	metre; milli	v	bottom function of a quotient; second function of a product; velocity; final velocity
m	gradient of a graph; mass		
M	mega		
M	molar mass	V	volt
mol	mole	V	volume
n	amount of material; number; number of electrons transferred in a redox reaction; number of X-ray reflections	V_m	molar volume $= V \div n$
		x	variable; the controlled variable; the horizontal axis on a graph ('abscissa'); mole fraction; displacement from equilibrium position
n_0	amount of material at the start of a process		
N	number of members in a series or data set; number of monomer units in a polymer; number of π-electrons	X	domain of an algebraic function
		y	variable; the observed variable; the vertical axis on a graph ('ordinate')
N_A	Avogadro number		
O	oxidized form of a redox couple	z	charge on an ion
p	pico	Z_eff	effective charge on an atomic nucleus
p	pressure; number of phases		
$p^{\ominus}$	standard pressure	γ	mean ionic activity coefficient
$P_{(x)}^{\ominus}$	partial pressure of pure compound x	δ	small increment; thickness of the Nernst diffusion layer
P	probability	Δ	Delta operator, as defined in eqn (2.1)
ph	phenyl		
pH	$-\log [\mathrm{H}^+]$	ε	molar extinction coefficient; permittivity
q	charge; heat	ε_0	permittivity of free space
Q	quinone	ζ	extent of reaction
Q	charge density; result of a Q-test	η	overpotential
$Q_{(\exp)}$	calculated Q-test quotient	θ	angle of focus in a triangle
r	correlation coefficient; radius of a circle or sphere	λ	wavelength
		Λ	ionic conductivity
R	reduced form of a redox couple	Λ^0	ionic conductivity at infinite dilution
R	gas constant; resistance; alkyl group		
		μ	reduced mass; ionic or electronic mobility
$\Re$	Rydberg constant		
s	second	ν	scan rate in cyclic voltammetry
s	sample standard deviation	π	the ratio of a circle's diameter and circumference
S	entropy		
t	time; temperature in Celsius	Π	Pi operator, as defined in eqn (2.4)
$t_{1/2}$	half-life	σ	population standard deviation

Σ	Sigma operator, as defined in eqns (2.2) and (2.3)
υ	viscosity
ϕ	electric potential
ψ	atomic wavefunction
ψ_{1s}	wavefunction of 1s atomic orbital
Ψ	molecular wavefunction
ω	angular rotation speed; frequency of a sinusoidal wave
Ω	ohm

Additional subscripts, superscripts, and typographic notations

$\ominus$	standard state
#	activation parameter
c	combustion
i	the ith member of a series
$\bar{x}$	an 'overbar' means a *mean* value of x, see eqn (21.1)

Standard prefixes

c	centi
d	deci
k	kilo
m	milli
M	mega
μ	micro
n	nano
p	pico

Acronyms and abbreviations

ac	alternating current
BODMAS	brackets, of, division, multiplication, addition, subtraction (page 29)
cos	cosine
DDT	dichlorodiphenyltrichloroethane
d.p.	decimal place
DRG	degree, radians, gradients (button on a calculator)
DVM	digital voltmeter
emf	electromotive force
IUPAC	International Union of Pure and Applied Chemistry
IQ	intelligence quotient
MV	methyl viologen (1,1′-dimethyl-4,4′-bipyridilium)
NMR	nuclear magnetic resonance
ppb	parts per billion
ppm	parts per million
RDE	rotated-disc electrode
SCE	saturated calomel electrode
s.f.	significant figure
sin	sine
SN	signal-to-noise ratio
sohcahtoh	acronym to use sin, cos, and tan (page 157)
tan	tangent

Irrational numbers

π(pi)	3.142...
τ(tau)	1.620...
e	2.178...

1

The display of numbers

Standard factors, scientific notation, and significant figures

Introducing algebraic phrases

By the end of this section, you will know:

- An algebraic phrase includes a **variable**, a **number**, and (usually in chemistry) **units**.
- One or more of the terms in the phrase may also be multiplied by a **factor**.
- We give the name **compound variable** to the product of two or more variables.

When chemists make a chemical compound, e.g. precipitating a salt by metathesis, they only require a certain number of building 'blocks'. The blocks in this case are anions and cations. Similarly, a surveyor tells the builder how many bricks and window frames are needed to build a house, and will write a quantity beside each on the order form: 10 000 bricks, 20 window frames, etc.

In a similar way, when we have a different **variable** such as velocity, we require the units of 'm' and 's^{-1}', and then quantify it, saying something like, 'He ran fast, covering a distance of 10 metres per second.' By this means, any variable is defined in terms of both a **number** and also its **units**. In chemistry, we generally write the variable with a symbol of some kind, enabling us to write an **algebraic phrase** such as energy $E = 12$ kJ mol^{-1}.

We write a phrase to describe the mass, length, velocity, etc., using a standard format:

$$\text{variable} = \text{number} \times \text{factor} \times \text{units} \tag{1.1}$$

> We give the name **compound unit** to several units written together. We need to leave a space between each constituent unit when we are writing such a compound, so kJ mol^{-1} is correct but kJmol^{-1} is not.

■ **Worked Example 1.1**

Look at the algebraic phrase 'energy $= 12$ kJ mol^{-1}'. Identify the variable, number, factor, and the unit.

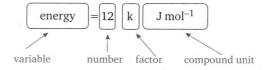

Reasoning:

- **Variable** In simple mathematical *phrases* such as this, we almost always write the variable on the left. A variable is a quantity whose value can be altered.
  - The variable on the left might be multiplied by other variables, in which case we have a *compound variable*.
- **Number** The easy part! It will be made up of numbers 1, 2, 3, ..., 0. The number could be written as a decimal or as a fraction.
- **Factor** If we need a factor, we always write it between the number and the unit(s).
  - Table 1.1 below is a comprehensive list of the standard factors, including micro µ, centi c, deci d, kilo k, mega M, etc. Alternatively, the factor could be written in terms of scientific notation.
- **Units** The units are always written on the far right of these phrases. In the example above, there are two units: joules (J) and moles (as mol^{-1}, in this case), so we call it a *compound unit*.
  - The way we write the unit(s) ought to follow the IUPAC scheme.
  - When there are several units, we separate each by a space.

A *factor* is simply a form of shorthand, so we could dispense with it by writing the *number* differently, saying energy $= 12\,000\,J\,mol^{-1}$, as below. By contrast, the units are *not* dispensable.

In the next sections, we will learn first how to write factors, then how to write the number in a meaningful way. ■

Self-test 1

In each of the following algebraic phrases, identify the variable, number, unit, and (where appropriate) the factor:

1.1	mass $= 2.65\,kg$		1.6	mass of large car $= 1.4\,Mg$
1.2	frequency $= 94.5\,MHz$		1.7	energy liberated $= 34\,MJ\,mol^{-1}$
1.3	wavelength $= 500\,nm$		1.8	length of bond $= 130\,pm$
1.4	current $= 0.3\,mA$		1.9	potential difference $= 550\,mV$
1.5	mass of beaker $= 340\,g$		1.10	speed $= 34\,m\,s^{-1}$

Writing algebraic phrases with a standard factor

By the end of this section, you will know:

- The standard factor is merely shorthand.
- The names, values, and symbols of the common IUPAC factors.
- How to write an algebraic phrase in terms of standard factors.

A typical chemical bond has a length l of about $0.000\,000\,000\,1\,m$ (where m here is the IUPAC symbol for 'metre', *not* milli). This number is cited in a ridiculous way: instinctively, we know that writing so many zeros not only is a waste of time and effort, but also increases the risk of error. For example, we could misread it and write either too many or too few. Surely there is a shorter way of relaying this information?

In this section, we look at a form of shorthand which is popular with chemists. Some numbers are huge and others tiny, but we never need to write out so many zeros, either before or after a decimal point.

For example, we could either say the length of a carbon–carbon double bond is $l = 0.000\,000\,000\,13\,m$, or abbreviate it by writing $l = 130\,pm$. Here, the small 'p' means 'pico', where $1\,pm = 0.000\,000\,000\,000\,1\,m$. We have not changed the length l in any way by saying it is 130 pm; we have only made it much easier to read or write.

The symbol 'p' is called a **standard factor** (or **prefix**). Table 1.1 lists the common standard factors. It is worth memorizing the factors in this table, because we will need them constantly. We never print these factors in italic type. Note how some letters are capitals and others are lower case (only factors greater than 1 appear in upper case).

> In chemistry, we give the name **factor** to any number by which we multiply the numerical value of a variable. Factors are usually expressed in shorthand notation.

> The name of the factor **pico** comes from the Latin *pica*, meaning 'tiny'. This root also helps explain the word 'piccolo' (i.e. a very small flute).

■ **Worked Example 1.2**

Burning 1 mole of *n*-octane (**I**), e.g. from petrol, liberates $5\,471\,000\,J$ of energy. Write this energy in terms of standard factors.

Table 1.1 Powers of 10: energy in joules.

10^{-18}	a	atto	0.000 000 000 000 000 001	1 aJ = energy of a single photon of blue light ($\lambda = 1 \times 10^{-7}$ m)
10^{-15}	f	femto	0.000 000 000 000 001	1 fJ = energy of a single high-energy photon (γ-ray of $\lambda = 10^{-10}$ m)
10^{-12}	p	pico	0.000 000 000 001	1 pJ = energy consumption of a single nerve impulse
10^{-9}	n	nano	0.000 000 001	1 nJ = energy per beat of a fly's wing
10^{-6}	μ	micro	0.000 001	1 μJ = energy released per second by a single phosphor on a TV screen
10^{-3}	m	milli	0.001	1 mJ = energy consumption per second of an LCD watch
10^{-2}	c	centi	0.01	1 cJ = energy per mole of low-energy photons (radio wave of $\lambda = 10$ m)
10^{-1}	d	deci	0.1	1 dJ = energy released by passing 1 mA across 1 V for 100 s (energy = $V\,i\,t$)
$10^0 = 1$	1		1	1 J = half the kinetic energy of 0.1 kg (1 N) travelling at a velocity of 1 m s^{-1}
10^3	k	kilo	1000	1 kJ = half the energy of room temperature (E at RT of 298 K = 2.3 kJ)
10^6	M	mega	1 000 000	1 MJ = energy of burning 1/3 mole of glucose (about a quarter of a chocolate bar)
10^9	G	giga	1 000 000 000	1 GJ = energy of 1 mole of γ-ray photons ($\lambda = 10^{-10}$ m)
10^{12}	T	tera	1 000 000 000 000	1 TJ = energy (via $E = mc^2$) held in a mass of 100 g
10^{15}	P	peta	1 000 000 000 000 000	1 PJ = energy released by detonating a very small nuclear bomb

In words, the energy released is about 5 million joules. The standard factor meaning 'million' is 'mega' (symbol M). We could write the energy as 5.471 × 1 000 000 J. By substituting for the factor of 'million' with the shorthand M, we obtain:

energy released = 5.471 MJ ■

The name of the factor **mega** comes from the Greek *megas* which means 'great'. Other common words coming from the same root include 'megaphone' (a device to make the voice bigger, i.e. louder) and 'megalith' (a huge stone such as those at Stonehenge).

Self-test 2

Rewrite the following using standard factors:

2.1 energy = 12 300 J mol^{-1}

2.2 frequency = 500 000 000 Hz

2.3 length of road = 3400 m

2.4 voltage = 30 000 V

2.5 mass of truck = 36 000 000 g

2.6 cost = £1 200 000

2.7 energetic content = 2 034 000 J

2.8 wavelength λ = 0.000 000 98 m

2.9 bond length = 0.000 000 000 156 m

2.10 diameter of a fly's eye = 0.000 01 m

Writing algebraic phrases using scientific notation

By the end of this section, you will know:

- Scientific notation is a different form of shorthand to the standard factors in the previous section, but fulfils the same function.
- How to write a number with standard factors as $a.b \times 10^c$.

Giga comes from the Latin *gigans*, meaning 'a giant'. We get the English word 'gigantic' from the same root.

The speed of light in a vacuum c is 300 000 000 m s^{-1}. We could write this speed with a standard factor, e.g. as $c = 0.3$ Gm s^{-1}, where G stands for giga (10^9).

Many people not only find it difficult to write a large number such as this, but also do not like writing the standard factors. They prefer using **scientific notation**. In this form of mathematical shorthand, we split the number into a simple two- or three-digit number (generally with a decimal point), followed by a factor. In this case, we avoid the standard factors in the previous section, but write the factor as 10 raised to an appropriate power. In the case of c above, we write the speed of light *in vacuo* as 3.0×10^8 m s^{-1}.

■ **Worked Example 1.3**

Write the Boltzmann constant k_B in scientific notation. The value of k_B is 0.000 000 000 000 000 000 000 013 8 J K^{-1}.

Aim:

Our aim is to write a number having the form $a.b \times 10^c$.

(i)   In this case, the number $a.b$ before the factor of 10^c will be simply 1.38.

(ii) We next look at the decimal place. Without the factor, the decimal point is 23 places to the left of the one, which is another way of saying 10^{-23}.

(iii) We then multiply the answers from parts (i) and (ii), to obtain 1.38×10^{-23}.

(iv) Finally, we usually need to include the units, which in this example are J K^{-1}.

Therefore, $k_B = 1.38 \times 10^{-23}$ J K^{-1}. ■

....

Aside

Using scientific notation on a calculator

We need to mention a common error experienced when typing numbers into a pocket calculator. The correct procedure when typing the Boltzmann constant would be:

(i)   Type '1.38'.

(ii) Press the button marked (EXP) or (EE) (which is usually positioned at the bottom of the number pad).

(iii) Type in '−23'.

It is *incorrect* to enter the number as:

(a) '1.38'

(b) $\times 10$

(c) Press (EXP) or (EE)

(d) Type in '23'.

<aside>
....
Typing '10' instead of '1' in step (b) here causes the answer to be 10 times too large.
....
</aside>

The calculator is programmed to think of 1.38×10^{-23} as $1.38 \times 1e^{-23}$, so typing in '10' in step (b) will introduce an additional factor of 10.

Using scientific notation on a computer spreadsheet

Using Excel^{TM}, we type '1.38e-23', where the small 'e' stands for 'exponent' (as explained in Chapter 10).

....

Self-test 3

3.1–10   Rewrite each of the algebraic phrases in Self-test 2 above, this time in terms of scientific notation.

3.11   The Faraday constant has a value of 96 500 C mol^{-1}. Rewrite this number in scientific notation.

3.12   The charge q on an electron is 0.000 000 000 000 000 000 16 C. Rewrite this charge in scientific notation.

The use of significant figures

By the end of this section, you will:

- Know what is meant by the term 'significant figures' (s.f.s).
- Appreciate that the number of significant figures is important.
- Appreciate how the number of significant figures makes a statement concerning the precision of our data: either we do not have the information available to allow greater precision, or the context means that more data are not useful.
- Know how to determine the correct number of significant figures.

Imagine a trip to a natural history museum, to look at the fossil of a great dinosaur. Beneath the bones is a small card saying '100 million years old' (i.e. an age of 100 000 000 years). Next, imagine going back again to see the fossil in the following year and on the same date. This second time, we notice how the display card has been amended slightly to read, '100 million and 1 years'. We would surely wonder how meaningful the additional 1 year was. The value of 100 million years is huge and therefore worth our attention; by contrast, citing an age as '100 million and 1' is silly. The 'and 1' tells us nothing when compared with the vastness of 100 million, because it is so small. The 100 million is significant; the 1 is simply irrelevant by comparison. In fact, we do not know the age precisely, so we signal that lack of knowledge by citing the age imprecisely.

In a different way, imagine two children describing their ages: the first says his age is five and a half years, but the second says her age is 6 years, 143 days and 3 hours. The first is giving an appropriate amount of information; the second gives us so much data that we would normally want to ignore most of it. It is foolish in this context to cite an age to this precision, because there is no need to know so much information.

In both these examples, dinosaurs and children, we see how the quantity of information says a lot about the precision of measurement. Often, particularly in everyday life, the precision with which a number is cited is erroneous, because the inferred precision is greater than the actual data allow. We need therefore to consider the correct way to cite numbers.

....
We often abbreviate the term **significant figures** with the letters **s.f.s**.
....

The number of **significant figures** is a commonly adopted measure of how precise or imprecise a number is. If we think it wise to express a lack of precision, we cite to only one s.f. If we have reason to feel more confident, we will use two s.f.s. And if we know our data are good, we may dare to cite to three or even four s.f.s.

■ **Worked Example 1.4**

The value of π is irrational. For most purposes, a value of 3.142 is sufficient. If we want an approximate value of π, we enter the quotient '22 ÷ 7' into a pocket calculator. The display says '3.142 857 140'.

How many significant figures are involved here?

- The value of π is 3 to **one** s.f.
- The value of π is 3.1 to **two** s.f.s.
- The value of π is 3.14 to **three** s.f.s.
- The value of π is 3.143 to **four** s.f.s.
- The value of π is 3.142 9 to **five** s.f.s.

- The value of π is 3.142 86 to **six** s.f.s.
- The value of π is 3.142 857 to **seven** s.f.s.
- The value of π is 3.142 857 1 to **eight** s.f.s.
- The value of π is 3.142 857 14 to **nine** s.f.s.
- The value of π is 3.142 857 140 to **ten** s.f.s.

Notice how we often need to round up or down, so, for example, while $\pi = 3.142\,86$ to six s.f.s, its value is 3.142 9 to five s.f.s.

In fact, the first few digits (3.14) are generally useful, and tell us the value of $22 \div 7$. The next three digits (286) are rarely worth citing: they do no more than 'tweak' our final answer. If we wanted the value of $\pi = 22 \div 7$ to calculate the area of a circle or the volume of a sphere or cone, the numbers '286' would not affect our answer.

Any further numbers (7140 ... etc.) are useless to just about everyone. They tell us nothing additional at all.

In summary, because we want only the first few significant figures of π, the quotient $22 \div 7$ generally gives a good enough value of 3.143, rather than 3.1429. ■

In practice, we usually calculate any parameter with slightly too many significant figures and then, as a final step, simplify by appropriately rounding up or down. For example, if we know our final answer should be expressed to three s.f.s, we perform the calculation using four s.f.s.

It is important to recognize that we cannot cite a quantitative answer without using significant figures. The number of significant figures makes a powerful statement about the precision of our method or experiment. For example, if we say $\Delta H = 12$ kJ mol^{-1}, we are in fact saying the technique used to measure ΔH was incapable of distinguishing between the enthalpy changes 11.5 and 12.5 kJ mol^{-1}. By contrast, if we say $\Delta H = 12.58$ kJ mol^{-1}, we imply the actual value lies between 12.575 and 12.585 kJ mol^{-1}. The latter experiment is 100 times more precise than the former.

> ....
> A number can be cited to any number of significant figures. *In chemistry, it is rarely useful to cite more than three s.f.s.*
> ....

Self-test 4

Re-express each of the following with a smaller number of significant figures; the required number of significant figures is given in square brackets after each algebraic phrase.

4.1   energy change $= -134.99$ kJ [3]

4.2 $emf = 1.4324$ V [4]

4.3   volume $= 1.986\,\mathrm{m}^3$ [2]

4.4   amount of material $= 3.221$ mol [3]

4.5   mass $= 1\,002\,010\,\mathrm{g}$ [1]

4.6   In SI units, a length of 1 metre $= 1\,650\,763.73$ wavelengths of the light emitted *in vacuo* by krypton-86 [7]

The use of significant figures when performing MULTIPLYING or DIVIDING

We often have to MULTIPLY or DIVIDE numbers expressed with differing numbers of significant figures. Again, we express the final answer rounded off to as many significant figures as the value with the least number of significant figures.

The old-fashioned name for the ester methyl salicylate is 'Oil of Wintergreen', and is commonly added to medications designed to alleviate muscular pain.

■ Worked Example 1.5

In a kinetics experiment, we wish to study the rate of forming the sweet-smelling ester methyl salicylate (**II**) from methanol (**III**) and salicylic acid (**IV**):

The rate of this reaction is given by the expression:

$$\text{rate} = k[\text{methanol}][\text{salicylic acid}]$$

where k is a rate constant. What is the value of the rate when $k = 3.06 \times 10^{-4}$ (mol dm^{-3})$^{-1}$s^{-1}, [methanol] $= 5$ mol dm^{-3}, and [salicylic acid] $= 0.45$ mol dm^{-3}?

These numbers are clearly expressed in terms of different numbers of significant figures. If we insert the numbers into the rate equation above, we obtain:

$$\text{rate} = 3.06 \times 10^{-4}(\text{mol dm}^{-3})^{-1}s^{-1} \times 5\,\text{mol dm}^{-3} \times 0.45\,\text{mol dm}^{-3}$$

so

$$\text{rate} = 6.885 \times 10^{-4} \text{ mol dm}^{-3}s^{-1}$$

The variable with the limiting precision was the concentration of methanol (which was expressed to one s.f.). Accordingly, we must express this rate to one s.f.:

$$\text{rate} = 7 \times 10^{-4} \text{ mol dm}^{-3}s^{-1} \quad ■$$

Self-test 5

Calculate the following MULTIPLICATION and DIVISION problems, citing each answer to the correct number of significant figures:

5.1 Volume of a unit cell, $V = $ side $a \times$ side $b \times$ side c. What is V if $a = 120$ pm, $b = 151$ pm, and $c = 146.5$ pm?

5.2 Amount of substance, $n = $ volume $\times$ concentration. What is n if the volume $= 0.250$ dm^3 and the concentration $= 0.05$ mol dm^{-3}?

5.3 Rate $= k \times$ [concentration]. Calculate the rate if $k = 9.3 \times 10^{-2}$ (mol dm^{-3})$^{-1}$ s^{-1} and [concentration] $= 0.3$ mol dm^{-3}.

5.4 Concentration, $c = $ amount of material $\div$ volume. Calculate c when the amount of material $= 3.2 \times 10^{-4}$ mol and the volume $= 0.5$ dm^3.

5.5 Amount of substance, $n = $ mass $\div$ molar mass. Calculate n when mass $= 4$ g and molar mass $= 422$ g mol^{-1}.

5.6 Charge density, $Q = $ charge $\div$ area. What is Q when charge $= 87.3$ mC and area $= 0.32$ cm^2?

5.7 From the ideal gas equation, $p = nRT \div V$. What is p if $n = 13$ mol, $R = 8.314$ J K^{-1} mol^{-1}, $T = 298$ K, and $V = 14.2$ m^3?

5.8 From the van't Hoff isotherm $\Delta G^{\ominus} = -RT \ln K$. What is the value of $\Delta G^{\ominus}$ if $\ln K = 4.0$, $T = 298$ K, and $R = 8.314$ J K^{-1} mol^{-1}?

The use of decimal places

By the end of this section, you will know:

- What is meant by the term 'decimal places'.
- How many decimal places to employ.
- The difference between decimal places and significant figures.

When we ADD or SUBTRACT numbers, we tend not to use significant figures. The results can be unhelpful or even ambiguous. Instead, we use **decimal places**. Decimal places are also called, confusingly, 'decimal *points*'. For example, the number 2.0 is expressed to one d.p. because there is a single digit after the dot. Conversely, the enthalpy change $\Delta H_c = 23.76$ kJ mol^{-1} is expressed to two d.p.s.

We use the correct number of decimal places to indicate the correct precision when ADDING or SUBTRACTING. The procedure is two-fold:

1. We look at each component term and note the number of decimal places in each, and note the minimum number of decimal places.
2. We ADD or SUBTRACT as normal, including *all* the decimal places from all the component terms. We will call this the 'preliminary' or 'uncorrected answer'.
3. We round the preliminary answer up or down to decrease the number of decimal places until it is the same as the value with the minimum number of decimal places.

■ Worked Example 1.6

Generally an electrochemical current I comprises several component parts. A current is measured and comprises the following three components: $I_{analyte} = 5.37$ mA, $I_{double\text{-}layer\ charging} = 43$ μA, and $I_{side\ reactions} = 1$ mA. What is the total current, expressed to an appropriate number of decimal places?

We start by writing a sum to enable the calculation of a total current:

$$I_{total} = I_{analyte} + I_{double\text{-}layer\ charging} + I_{side\ reactions}$$

where $I_{analyte}$ is expressed to two d.p.s, $I_{double\text{-}layer\ charging}$ is expressed to three d.p.s, and $I_{side\ reactions}$ is expressed to zero d.p.s. Therefore we must express the final answer to zero d.p.s.

Before we calculate the final sum, it is usually helpful to express the numbers in a consistent way, with each component having the same common factor. That is:

$$I_{total} = 5.37\text{mA} + 0.043\text{mA} + 1\text{mA}$$

so:

$$I_{total} = 6.413\ \text{mA}$$

When expressed with a precision of zero d.p.s, this current is 6 mA.

Occasionally, when the numbers are *not* comparable in magnitude, we will sometimes exclude one or more component parts because they are too small. We then express the final answer to the minimum number of decimal places of the remaining components. ■

....
We tend to 'exhaust the data' when we subtract two nearly equal numbers, so it may be wise to perform the calculation with too many decimal places, then redisplay the answer.
....

■ **Worked Example 1.7**

An industrial chemist wishes to make a batch of paint. The paint is to be made in a vat. To this vat, the chemist adds 12.0 tonnes of titanium dioxide (as a 'filler'), 5.0 tonnes of polyurethane monomer, and 15 kg of pigment.

Cite the overall mass to an appropriate number of decimal places.

To ensure consistency, it is convenient first express each mass with the same units. In this case, expressed in tonnes, the mass of pigment is 0.015 tonnes. We obtain:

$$12.0 + 5.0 + 0.015 = 17.015$$

The third mass is clearly inconsequential when compared with the others, so we simply omit it. Such omission is the same as citing the answer to one d.p.:

overall mass $= 12.0 + 5.0$ tonnes $= 17.0$ tonnes

It would not have made sense to say '17.015 tonnes' (which involves three d.p.s) because the two larger masses are each expressed to only one d.p. In other words, the spread of values expressed by the next smallest mass '5.0 tonnes' is itself greater than the mass of 15.0 kg. ■

Self-test 6

Calculate the following ADDITION and SUBTRACTION problems, citing each answer to the correct number of decimal places:

6.1 mass $= 12.0$ g $+ 1.001$ kg $- 130.62$ g

6.2 charge $= 96\,000$ C $- 67.27$ C $- 1096.3$ C

6.3 amount of material $= 12.1$ mol $- 2.754$ mol $+ 0.5419$ mol

6.4 time $= 60.4$ s $+ 12$ μs $+ 33.96$ ms $+ 4.0$ s

Additional problems

1.1 The mass of a proton is 1.66×10^{-27} kg. Rewrite this figure with a standard factor.

1.2 Cobalt chloride hexahydrate has a molar mass of 220 g mol^{-1}. A sample of mass 1.5 g is dissolved in water (140 cm^3). What is the concentration of the solution, cited to one, two, and three s.f.s?

1.3 A protein has a molar mass of 1.302×10^5 g. Write this mass in the form of a straightforward number.

1.4 The current i through an electrode is 23.4 μA. The area A of the electrode is 4.1×10^{-4} m^2. What is the current density I, where $i = I \div A$?

1.5 The closest distance between two chloride nuclei in a crystal of sodium chloride is 0.000 000 000 174 m. Write this length in terms of scientific notation.

1.6 The Faraday constant F is obtained as the product of the Avogadro number $N_A = 6.022 \times 10^{23}$ mol^{-1} and the charge on a single electron $q = 1.602 \times 10^{-19}$ C, i.e. $F = N_A \times q$. Calculate the value of F.

1.7 The velocity of light c is 2.9980×10^8 m s^{-1}. Write this velocity to two s.f.s.

1.8 The gradient of a graph is defined as the change in the vertical axis Δy divided by the change in the horizontal axis Δx, i.e. gradient $= \Delta y \div \Delta x$. Calculate the gradient if $\Delta y = 0.000\,324$ and $\Delta x = 41$.

1.9 The number of transistors produced worldwide each year is 2×10^{19}. Write this number with a standard factor.

1.10 A chemist makes 0.37 g of potassium chloride. The molar mass of KCl is 74.5 g mol^{-1}. If 'amount = mass ÷ molar mass', what is the amount of KCl made?

2

Algebra I

Introducing notation, nomenclature, symbols, and operators

Mathematical shorthand: symbols

By the end of this section, you will know:

- Mathematical symbols are merely shorthand, each telling us to *do* something.
- All mathematical expressions contain operators.

Our word 'mathematics' comes from the Greek *mathatās*, which means a 'disciple'. A mathematician is therefore someone who closely follows the dictates of a master—in this case, the simple rules of how mathematical symbols behave.

We already know many mathematical symbols, such as $+$, $-$, $\times$, $\div$, and so on. We must first appreciate how these symbols are merely a form of **shorthand**, intended to save us time and ink. They are never more than **symbols**, and are never magical.

The simplest symbol is the plus sign. Each plus sign '$+$' we encounter operates in the same way: it tells us to do something; in effect, it says, 'look at the numbers either side of me and add them together'. For this reason, we call the plus sign an **operator**, because it tells us what operation to perform on the adjacent numbers. In the same way, a multiplication sign '$\times$' is also an operator, because it tells us to multiply together whatever is positioned to its immediate left and immediate right.

While these examples are utterly trivial, they tell us a lot about the way that even a simple operator works, and demonstrate how they are merely one form of shorthand notation. In each case we could have used words. We have a choice, and could say either 'the number obtained when we assemble eight groups of five' or '8×5'. We obtain the same answer in both cases: the only difference is the *notation*.

In summary, we use mathematical operators to save time and space, not to make life more complicated. The rest of this course merely explores the mathematics of operators, and how chemists can master them.

> We call the plus sign an **operator**, because it tells us what operation to perform.

> An operator is merely a shorthand notation.

> The word **notation** means the way we write something by using symbols, especially in music, linguistics, or mathematics.

The Delta operator Δ

By the end of this section, you will know that the Delta operator Δ:

- Is a shorthand symbol.
- Is meaningless written on its own: we write it with a **domain**.
- Tells us to subtract the initial value of the domain from the final value of the domain.

As chemists, we often see the term ΔH, particularly when we study thermodynamics and calorimetry. As we learn more chemistry, we will also encounter ΔS, ΔU, and ΔG. In each case, the symbol Δ is the Greek letter **Delta**. (We write 'Delta' rather than 'delta', because a lower-case delta looks like δ.)

The **phrase** ΔH is the change in enthalpy during a reaction: enthalpy has the symbol H, so the symbol Δ must be some form of shorthand for 'change in enthalpy'. ΔH therefore does not mean 'Delta multiplied by H'; in fact, if we want to gain the full significance of Δ, we ought to say 'Delta of H'. The symbol Δ is intended to tell us that *something happens* to H. Because Δ does something to H, we say that

Δ **operates** on H. In more technical language, Δ is an **operator**. If we were to see the symbol Δ on its own, it would not be an operator: *an operator must operate on something*. In fact, a symbol means what we want it to mean: we say that we **define** its meaning. In other words, the only reason Δ means what it means is because we say it does. Table 2.1 contains a list of other symbols, some of which are also operators. We will need to learn this list because there is no inner *logic* here—for example, there is nothing about a capital H that makes us say 'entrophy'.

So, as an operator, Δ must operate *on* something. We call that something the **domain**. The word 'domain' usually means something spatial—the area controlled by a king or government; but in mathematics, 'domain' means the thing over which the operator has control. In the case of the phrase ΔH, the operator is Δ and the domain is H. We will investigate the scope of a domain as we progress through this chapter.

The Δ operator tells us to subtract the initial value of the domain from the final value of the domain. For a domain in general (call it X):

$$\Delta X = X_{final} - X_{initial} \qquad (2.1)$$

so when we see the phrase ΔH, we are to subtract the initial value of enthalpy H from the final value. We can show this diagrammatically in Fig. 2.1.

It will be clear from Fig. 2.1 that the value of ΔH has not only a magnitude—the spatial separation between the two bold lines—but also a sign.

> It is vitally important to distinguish between Δ and H : Δ is an **operator** because it *does something* (according to eqn (2.1)). H is a **symbol** because it is *shorthand* for enthalpy.

Table 2.1 List of symbols having more than meaning.

Symbol	Meaning
Δ	difference, according to eqn (2.1); crystal-field-splitting energy
σ	standard deviation; electronic conductivity
c	concentration; constant; intercept on a graph
I	electrical current; intensity; ionic strength
k	rate constant; kilo; constant
K	equilibrium constant; potassium; kelvin
T	transmittance; tritium
U	internal energy; uranium
x	horizontal axis on a graph; unknown; mole fraction

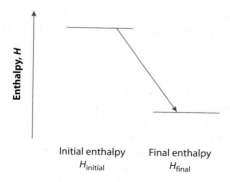

Figure 2.1 In an **exothermic** process, the final product has *less* energy than the initial starting materials. Energy has been given out.

■ **Worked Example 2.1**

We give the letter S to entropy, so ΔS is the change in entropy. If $S_{final} = 12$ J K^{-1}mol^{-1} and $S_{initial} = 25$ J K^{-1}mol^{-1}, calculate ΔS.
Inserting values into eqn (2.1):

$$\Delta S = S_{final} - S_{initial}$$

$$\Delta S = (12 - 25) \text{ J K}^{-1}\text{mol}^{-1}$$

$$\Delta S = -13 \text{ J K}^{-1}\text{mol}^{-1}$$

So we calculate the value of ΔS as -13 J K^{-1}mol^{-1}. The change in S is negative. ■

Self-test 1

1.1 If the final value of X is 10 and the initial value is 22, what is ΔX?

1.2 If the final value of X is 33 and the initial value is 5.2, what is ΔX?

1.3 If the initial value of X is 9.34 and the final value is 9.37, what is ΔX?

1.4 If the final value of the Gibbs function G is 8.0 kJ mol^{-1} and the initial value of G is 3.6 kJ mol^{-1}, what is ΔG?

1.5 If the optical absorbance Abs decreases from 1.1 to 0.6 during a reaction, what is the value of $\Delta(Abs)$?

1.6 If the electromotive force emf increases from 1.45 V to 1.50 V during a process, what is the value of $\Delta(emf)$?

The Sigma operator Σ

By the end of this section, you will know that:

- The Sigma operator Σ is a shorthand symbol.
- The Sigma operator Σ tells us to add up a series of terms.
- The terms to be summed can themselves be sums of smaller terms.

■ **Worked Example 2.2**

A chemist is stocking the laboratory before starting an experiment and needs to order conical flasks. The first part requires 2 flasks, the second 15, and the third 4. The final part requires 3 flasks. What is the overall number required?

To perform this trivial calculation, we merely say

$$\text{cost} = (2 \text{ flasks}) + (15 \text{ flasks}) + (4 \text{ flasks}) + (3 \text{ flasks})$$

which obviously amounts to 24 flasks.

It is possible to write this calculation in a form of shorthand: we introduce the Greek symbol sigma Σ, which means 'the sum of'. We write:

$$\sum i = \text{1st item} + \text{2nd item} + \cdots + \text{last item} \tag{2.2}$$

In this case, we could then write:

$$\text{total number} = \sum(\text{flasks})$$

In words, this assembly of symbols says, 'The total number of flasks amounts to the sum of all the individual numbers of flasks.'

More usually, we amend this notation somewhat, and write eqn (2.2) as:

$$\text{total number} = \sum_{i=1}^{n}(\text{flasks})_i \qquad (2.3)$$

where the small 'i' beneath the Σ means 'per item'. In this trivial example, each item had the same value—it is a flask—but such situations are rare. The term '$i=1$' in small letters beneath the Σ and the 'n' above it indicate the scope of the sum: it tells us to sum all the costs from the first ($i=1$) to the last ($i=n$). In other words, there are n samples. We could therefore write:

$$\text{cost} = \sum_{i=1}^{n}(\text{flasks})_i = \text{flasks}_{\text{experiment}_1} + \text{flasks}_{\text{experiment}_2}$$

$$+ \text{flasks}_{\text{experiment}_3} + \text{flasks}_{\text{experiment}_4}$$

The value of n is 4 in Worked example 2.2 because there are four terms. ■

■ **Worked Example 2.3**

What is the mass per mole of the ozone-depleting compound fluorochloroiodo methane (**I**)?

Strategy:

(i)   We write a sum using the sigma notation in the style of eqn (2.3).

(ii)   We then add together the mass of the atoms.

(i)   The overall mass $= \sum_{i=1}^{5}(\text{atoms})_i$.

The molecule (**I**) has five atoms (hence the number on top of the Σ): hydrogen, carbon, fluorine, chlorine, and iodine. The relative masses per atom are respectively 1, 12, 19, 35.5, and 127.

(ii)   The overall mass is therefore

$$\sum_{i=1}^{n}(1 + 12 + 19 + 35.5 + 127).$$

We see how the overall mass is 194.5. ■

This example is very simple. At first sight it looks as though Σ is merely a new way of writing a straightforward plus sign, $+$. In fact, we generally write a sum with a sigma symbol Σ when we want to add up several smaller sums. For example, we write a new version of eqn (2.3):

$$\text{mass per molecule} = \sum_{i=1}^{n}[(\text{number of atoms of element } i)$$

$$\times (\text{mass per atom of element } i)]$$

where the brackets indicate the scope of individual terms in the sum.

We employ a similar calculation to add up the molar mass for more molecules than (**I**): for each element in the molecule, we multiply together the number of atoms of that element within the molecule and the atomic mass of that element. Each little sum yields the mass of (say) carbon, hydrogen, or oxygen present in 1 mole of the molecule. We then add up the total mass of all of the elements to obtain the molar mass of the molecule (the mass of 1 mole of the molecule). ■

■ Worked Example 2.4

Calculate the mass per mole of the amino acid L-proline (**II**).

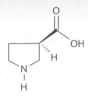

II

We first write the elemental formula of L-proline as $C_5H_9O_2N_1$. We then calculate the mass of 1 mole. We start by writing a line of sums in the style of eqn (2.3):

$$\text{mass per mole of (II)} = \sum_{i=1}^{n} [(\text{number of atoms of type } i) \times (\text{mass per atom of type } i)]$$

so:

> We include the units $g \, mol^{-1}$ here because they are the SI units of molar mass.

$$\text{mass of 1 mole of (II)} = (C \; 5 \times 12 \; g \, mol^{-1}) + (O \; 2 \times 16 \; g \, mol^{-1}) \\ + (N \; 1 \times 14 \; g \, mol^{-1})$$

$$\text{mass of 1 mole of (I)} = (C \; 60 \; g \, mol^{-1}) + (H \; 9 \; g \, mol^{-1}) + (O \; 32 \; g \, mol^{-1}) \\ + (N \; 14 \; g \, mol^{-1})$$

$$\text{mass of 1 mole of (I)} = 115 \; g \, mol^{-1} \quad ■$$

■ Worked Example 2.5

We can calculate the standard enthalpy change of reaction $\Delta H_r^{\ominus}$ as a sum of the molar enthalpies of formation $\Delta H_f^{\ominus}$ for each chemical participating in a reaction:

$$\Delta H_r^{\ominus} = \sum_{products} (v \, \Delta H_f^{\ominus}) - \sum_{reactants} (v \, \Delta H_f^{\ominus})$$

where v here is the **stoichiometric number**: that is, the number before the participating species in a fully balanced chemical reaction.

Consider the oxidation reaction occurring in a catalytic converter:

$$2CO_{(g)} + O_{2(g)} \rightarrow 2CO_{2(g)}$$

Calculate a value of $\Delta H_r^{\ominus}$ for this process. Take $\Delta H_f^{\ominus}(CO) = -110.5 \; kJ \, mol^{-1}$, $\Delta H_f^{\ominus}(O_2) = 0$, and $\Delta H_f^{\ominus}(CO_2) = -393.0 \; kJ \, mol^{-1}$.

Firstly, from the balanced reaction, the respective values of v are therefore:

$$v_{(CO)} = 2, v_{(O_2)} = 1, \text{ and } v_{(CO_2)} = 2$$

Secondly, we write an expression based on the template of eqn (2.2), saying:

$$\Delta H_r^{\ominus} = 2 \times \Delta H_f^{\ominus}(CO_2) - (2 \times \Delta H_f^{\ominus}(CO) + 1 \times \Delta H_f^{\ominus}(O_2))$$

Inserting values:

$$\Delta H_r^{\ominus} = 2 \times -393.0 \text{ kJ mol}^{-1} - (2 \times (-110.5 \text{ kJ mol}^{-1}) + 1 \times (0))$$

so:

$$\Delta H_r^{\ominus} = -786.0 \text{ kJ mol}^{-1} - (-221.0 \text{ kJ mol}^{-1})$$

and

$$\Delta H_r^{\ominus} = -565 \text{ kJ mol}^{-1} \quad \blacksquare$$

Self-test 2

For each of the following, starting with an expression involving a summation sign Σ add up the molecular mass:

2.1 Potassium hexacyanoferrate(II), $K_4[Fe(CN)_6].3H_2O$

2.2 Anhydrous copper sulphate, $CuSO_4$

2.3 Hydrogen peroxide, H_2O_2

2.4 Benzoic acid, $C_7H_6O_2$ **(III)**

2.5 Pyrrole, $C_5N_1H_5$ **(IV)**

2.6 Aspirin, $C_9O_4H_8$ **(V)**

| III | IV | V |

Take the relative atomic masses as $H = 1$, $C = 12$, $N = 14$, $O = 16$, $S = 32$, $K = 39$, $Fe = 56$, and $Cu = 64$.

We can even use this approach to calculate **the sum of a sum**. So, for example, we can calculate the molar masses of reactants, and then sum them. In this case, we write:

$$\text{mass of product} = \sum (\text{mass per reactant})$$

Here, 'mass per reactant' is itself a sum:

$$\text{mass of reactant} = \sum (\text{number of respective atoms} \times \text{mass per atom})$$

Some people might wish to combine these two sums, writing each in the style of eqn (2.2):

$$\text{mass of product} = \sum \left(\text{masses of} \left[\sum (\text{number of atoms} \times \text{mass per atom})\right]\right)$$

or even:

$$\text{mass of product} = \sum_{i=1}^{n}\left(\sum_{j=1}^{n} X_j\right)$$

Looking closely beneath each sigma symbol, we see one has a small i and the other a small j. This notation merely emphasizes that each summation adds up a different set of numbers.

■ **Worked Example 2.6**

In a solution of methanol (**VI**), methyl viologen (MV) (**VII**) and phenol (PhOH) (**VIII**) join together to form a molecular complex:

$$2CH_3OH + MV + 2PhOH \rightarrow \text{complex}$$

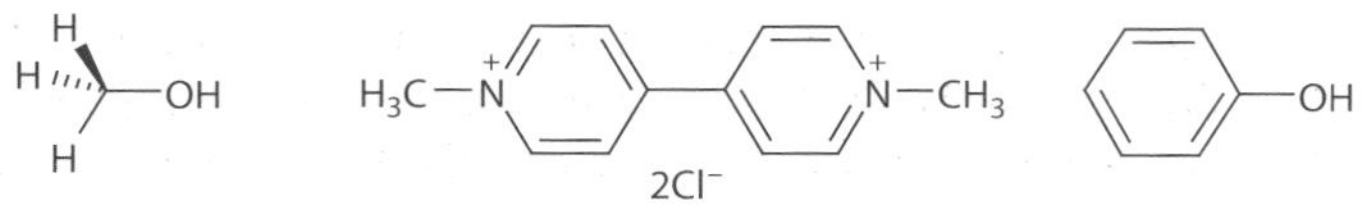

VI **VII** **VIII**

Calculate the mass per mole of the complex produced.

We first perform three sums to calculate the mass per mole of (**VI**), (**VII**), and (**VIII**):

$$\text{mass of } (\mathbf{VI}) = \sum(\text{C: } 1 \times 12) + (\text{H: } 4 \times 1) + (\text{O: } 1 \times 16) = 32\,\text{g mol}^{-1}$$

$$\text{mass of } (\mathbf{VII}) = \sum(\text{C: } 24 \times 12) + (\text{H: } 4 \times 1) + (\text{N: } 2 \times 14) + (\text{Cl: } 2 \times 35.5)$$
$$= 401\,\text{g mol}^{-1}$$

$$\text{mass of } (\mathbf{VII}) = \sum(\text{C: } 6 \times 12) + (\text{H: } 6 \times 1) + (\text{O: } 1 \times 16) = 94\text{ g mol}^{-1}$$

Next, we calculate the mass of complex produced, saying:

$$\begin{matrix}\text{molar mass}\\ \text{of complex}\end{matrix} = \sum(\text{molar masses of constituent molecules})_i$$

so:

$$\begin{matrix}\text{molar mass}\\ \text{of complex}\end{matrix} = \sum(2 \times \text{mass of } (\mathbf{VI})) + (1 \times \text{mass of } (\mathbf{VII})) + (2 \times \text{mass of } (\mathbf{VIII}))$$

and:

$$\begin{matrix}\text{molar mass}\\ \text{of complex}\end{matrix} = \sum(2 \times 32) + (1 \times 401) + (2 \times 94)$$

We see how the mass per molar of the complex is $653\,\text{g mol}^{-1}$. ■

Self-test 3

3.1 Consider the reaction between acetone (**IX**) and 1,2-dihydroxyethane (**X**) to form a cyclic acetal.

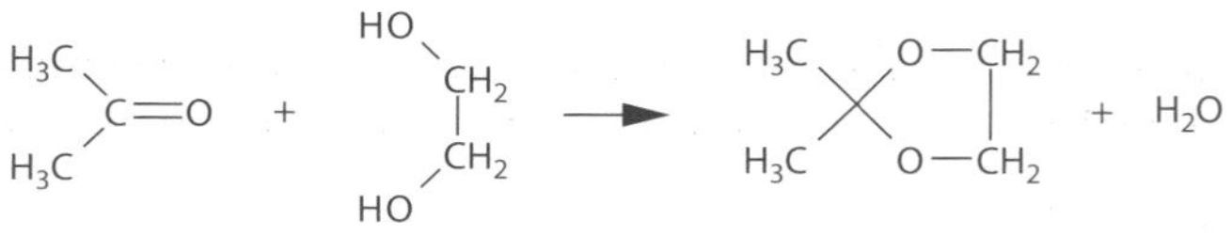

IX **X**

Scheme 2.1

Assuming the reaction is quantitative, what is the mass of product if we react 0.35 mol each of (**IX**) and (**X**)?

Formulate the answer in the form of eqn (2.3).

Continued...

⟩⟩ *Self-test 3 Continued* . . .

3.2   A student's overall mark depends on the percentage of each constituent unit studied. Calculate the student's overall mark if:

  (i)   The maths unit is worth 10 credits, and has a percentage of 45.
  (ii) The inorganic unit is worth 20 credits, and has a percentage of 56.
  (iii) The organic unit is worth 20 credits, and has a percentage of 70.
  (iv) The physical unit is worth 20 credits, and has a percentage of 62.
  (v)   The laboratory unit is worth 30 credits, and has a percentage of 40.

The Pi operator Π

By the end of this section, you will know that the Pi operator Π:

- Is a shorthand symbol.
- Tells us to multiply together a series of terms.

In everyday life, we could say 'multiplying together the numbers 4 and 5 yields the number 20' or 'the product of 4 and 5 is 20'. Both are correct, but the latter is shorter.

The word **product** has the mathematical sense of 'multiply together'. A **Pi product** employs is another shorthand operator. (The word 'product' is again used in its mathematical context.) In terms of usage, a Pi product is an expression starting with the Greek symbol Pi (Π), used as a shorthand way of saying, 'multiply together series of terms'. We call it Pi with a capital letter, because a lower-case pi looks like π.

> The word **product** here is a mathematical term meaning 'multiply together'.

■ **Worked Example 2.7**

What is the Pi product of the numbers 4, 6, and 9?
The value of the Pi product is $4 \times 6 \times 9 = 216$.

When we write a Pi product (sometimes typed as Π product) we follow the Π symbol with a phrase or bracket. In mathematical notation, we would write the numbers after the Π symbol, each number or item separated by a comma:

$$\prod(4, 6, 9) = 216$$

Occasionally, we need to multiply together a series of generic terms. In this case, we tend to write the Pi product in a slightly more complicated-looking way:

$$\prod_{i=1}^{n} X_i$$

> The word **generic** here means 'concerning origin' and comes from the Greek word *genesis*, itself meaning 'in the beginning'. The English *genetics* comes from this same root.

where the term X_i means a mathematical description of the general species i. There are n terms to be multiplied together. The subscripted phrase '$i = 1$' means that we start with the first, and the superscripted 'n' tells us that we continue through to the last, nth term. We could have written this definition above as eqn (2.4):

$$\prod_{i=1}^{n} X_i = X_1 \times X_2 \times X_3 \times \cdots \times X_n \tag{2.4}$$

■

■ Worked Example 2.8

A common way of denoting an equilibrium constant K is:

$$K = \frac{\prod [\text{products}]}{\prod [\text{reactants}]}$$

where the square brackets indicate concentrations. Write an expression for the equilibrium constant K for the oxidation reaction:

$$Ce^{IV} + Fe^{II} \rightarrow Ce^{III} + Fe^{III}$$

The products are written on the right-hand side and the reactants on the left. We write the equation therefore as:

$$K = \frac{\prod [Ce^{III}], [Fe^{III}]}{\prod [Ce^{IV}], [Fe^{II}]}$$

or, in a more familiar notation, as:

$$K = \frac{[Ce^{III}] \times [Fe^{III}]}{[Ce^{IV}] \times [Fe^{II}]}$$

In fact, most people omit the multiplication signs, as:

$$K = \frac{[Ce^{III}][Fe^{III}]}{[Ce^{IV}][Fe^{II}]}$$

All three notations are correct and valid. ■

Sometimes we write equilibrium constants in such a way that we retain the word 'product'.

We need to be aware how often there will be different, but equally valid, notations.

■ Worked Example 2.9

Solid limestone ($CaCO_3$) is almost water insoluble. The process of dissolving limestone is:

$$CaCO_{3(s)} \rightarrow Ca^{2+}_{(aq)} + CO^{2-}_{3(aq)}$$

Write the equilibrium constant K for this process and hence work out the relationship between the equilibrium constant and a Pi product. (Do not include the $CaCO_3$ in the equilibrium constant K, because it is a solid.)

We write the equilibrium constant K for this process saying:

$$K = [Ca^{2+}_{(aq)}] \times [CO^{2-}_{3(aq)}]$$

The right-hand side is a Pi product, so we write this simplified equilibrium constant as:

$$K = \prod (\text{concentrations of ions in solution})_i \quad ■$$

We often call this equation 'the **solubility product** of limestone', and give K the additional subscript 'sp', as K_{sp}. The word 'product' here alerts us to the need to multiply together the ionic concentrations.

■ Worked Example 2.10

A chemist wishes to make the complexing ligand 4-(4-bromophenyl)-2,2'-bipyridine (**XI**) in three steps, according to Scheme 2.2. The yield of step **1** is 66%, the yield of step **2** is 92%, but the yield of step **3** is only 41%. What is the overall yield of (**XI**)?

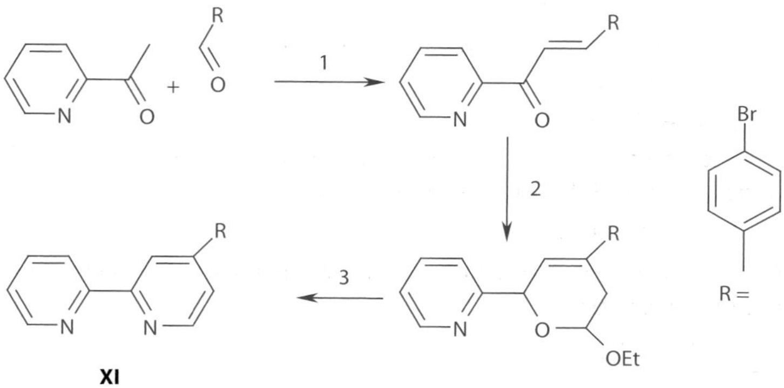

Scheme 2.2

We calculate the overall yield of (**XI**) as a Pi product in the style of eqn (2.4), saying:

$$\text{overall yield} = \prod_{i=1}^{3}(\text{yield of step } i)$$

so we compute by multiplying together the individual yields:

overall yield of (XI) = (yield of step 1) × (yield of step 2)
× (yield of step 3)

so:

overall yield of (XI) = $(0.66) \times (0.92) \times (0.41)$

and:

overall yield of (XI) = 0.249 or 24.9%

The rather low overall yield reminds us that, when we perform a multi-step reaction, we need to maximize the yield of each step if we are to generate an economically viable amount of product. ■

Sometimes we wish to write a Pi-product expression from a series of numbers.

....
Note how we talk about yields in terms of **percentages**. In reality, a percentage is merely a *proportion* multiplied by 100, i.e. the proportion with the decimal point repositioned.
....

■ **Worked Example 2.11**

Consider the following series of numbers:

$$1 \times 4 \times 9 \times 16 \times 25 \times 36$$

Write a Pi-product expression in the style of eqn (2.4) to describe them.

A quick glace tells us we have a series of square numbers, ranging from 1^2 to 6^2.

Strategy:

(i)   We write a letter Pi, because the members of the series are multiplied together: Π

(ii)   The variable i has been squared, so we write against the Π the term 'i^2': $\Pi\ i^2$.

(iii)   The lowest value of i in the series is 1, so we write '$i = 1$' beneath the Π term: $\prod_{i=1} i^2$

(iv)   Finally, the highest value of i in the series is 6, so we write '6' above the Π term: $\prod_{i=1}^{6} i^2$. ■

....
In writing this expression, it is assumed that values of i are **integers**, i.e. whole numbers.
....

Other operators commonly encountered in chemistry

We have looked at three of the chemist's most common operators, Delta, Sigma, and Pi. They have much in common: firstly, we write a symbol for each using a capital Greek letter. More importantly, they save us a lot of time because they allow us to represent a mathematical reality using a form of shorthand notation.

Table 2.2 contains a series of other common operators together with their notations.

We need to note the following:

- The operator is generally written to the *left* of the domain. We write $\sin x$ rather than $x \sin$. Power operators such as square, cube, etc., are only common exceptions to this rule. (We describe powers in Chapter 10.)

- Some operators have more than one correct notation. For example, $\sqrt{X}$ or $\sqrt{X}$ are equally valid; and $\exp(X)$ and e^X are also equally valid. (We describe exponentials in Chapter 11.)

- Each time we write a domain X, we must recall it could be a constant number (whole or partial), or a variable of some kind. The variable can be simple, such as a single term (velocity, mass, current), or a series of variables arranged in an equation. Therefore, if we write an expression such as X^2, the domain X could be 2, 3.373, velocity, or even a complicated string, such as $(x^2 - \sqrt{x} + (2/x))$.

- If we write $\sqrt{X}$ (with a bar positioned over the X), we are being told to take the root of everything enclosed within the root symbol, $\sqrt{}$. By contrast, writing, $\sqrt{X}$ could mean the root of *everything* positioned to the right of the $\sqrt{}$ symbol—a situation that is frequently ambiguous. Therefore we should enclose the domain within brackets if it comprises more than a single term. As an example, we write $\sin(x^3)$ rather than $\sin x^3$, because $\sin x^3$ could mean the different function $(\sin x)^3$. Similarly, we write $(2\pi f)^2$ if the domain is $2\pi f$, because $2\pi f^2$ could mean merely $2 \times \pi \times f \times f$, with the 2 and the π not being squared. (We describe trigonometric functions such as sine in Chapter 13.) In this context, a root sign can be considered to be a bracket, so writing $\sqrt{(mgh)}$ is the same as writing $\sqrt{mgh}$

Table 2.2 Operators commonly encountered in chemistry. In each case here, the letter X represents the domain.

Function	Symbol	Usual notation	Simple example
square	2	X^2	$3^2 = 3 \times 3 = 9$
cube	3	X^3	$4^3 = 4 \times 4 \times 4 = 64$
powers (general)	n	X^n	e.g. if $n = 3$, $3^3 = 3 \times 3 \times 3 = 27$
square root	$\sqrt{}$	$\sqrt{X}$ or $\sqrt{} X$	$\sqrt{4} = 2$
cube root	$\sqrt[3]{}$	$\sqrt[3]{X}$ or $\sqrt[3]{} X$	$\sqrt[3]{8} = 2$
percentage	%	$X\%$	$= (\text{proportion of } X) \times 100$
sine	sin	$\sin X$ or $\sin (X)$	
cosine	cos	$\cos X$ or $\sin (X)$	
logarithm in base 10	log	$\log X$ or $\log_{10} X$, or $\log (X)$ or $\log_{10}(X)$	
logarithm in base e	ln	$\ln X$ or $\ln_e X$ or $\ln_e(X)$	

Additional problems

2.1 A chemist weights out a mass m. The balance display indicates the value of m is 3262 g. Use this information to write an algebraic phrase in terms of kilograms.

2.2 Consider the expression 'current $= 35 \ \mu$A'. Identify each part of this algebraic phrase.

2.3 When plotting a graph, the vertical axis (the y axis) increases from 12 cm to 32 cm. What is the value of Δy?

2.4 Determine a value of the sum $\sum$ (1, 1.1, 1.4, 1.5, 6.2).

2.5 To obtain the standard **enthalpy change on combustion** $\Delta H_c^{\ominus}$, we sum the molar enthalpies of each chemical participating in the reaction:

$$\Delta H_c^{\ominus} = \sum_{\text{products}} v H_f^{\ominus} - \sum_{\text{reactants}} v H_f^{\ominus}$$

Calculate the value of $\Delta H_c^{\ominus}$ for methane at 25°C using the molar enthalpies of formation, $\Delta H_f^{\ominus}$. The necessary data are:

Species (all as gases)	CH_4	O_2	CO_2	H_2O
$\Delta H_f^{\ominus} / \text{kJ mol}^{-1}$	-74.81	0	-393.51	-285.83
v	1	2	1	2

The stoichiometry of the reaction is: $CH_4 + 2O_2 \rightarrow CO_2 + 2H_2O$.

2.6 One definition of **equilibrium constant** is:

$$K = \frac{\prod [\text{products}]}{\prod [\text{reactants}]}$$

Now consider the following redox reaction:

$$Ce^{IV} + Co^{II} \rightarrow Ce^{III} + Co^{III}$$

Calculate a value of K for this reaction using the following equilibrium concentrations: $[Ce^{IV}] = 3.22 \times 10^{-9}$, $[Co^{II}] = 2.9 \times 10^{-8}$, $[Ce^{III}] = 0.09$, and $[Co^{III}] = 0.4361$, all expressed in the same units of mol dm^{-3}.

2.7 Determine a value of the product $\prod (1.2,4,5.5,7,9)$.

2.8 Determine a value of the product $\prod_{i=3}^{5}(i^2/3)$.

2.9 During the course of a reaction, the entropy S changes from -15.1 J K^{-1}mol^{-1} to -32.5 J K^{-1}mol^{-1}. Calculate a value of ΔS.

2.10 Determine a value of the sum $\sum_{i=2}^{5} i^2$.

3

Algebra II

The correct order to perform a series of operations: BODMAS

The correct order to perform a calculation: BODMAS

By the end of this section, you will know:

- When a calculation involves more than one operator, the resultant value depends on the order in which we use the operators.
- Only if we perform the operations in the right order will the answer be correct.
- The BODMAS scheme is a simple way to remember the correct order in which to perform operations.

When we write a sentence, we need to be careful that the words say what we want them to say. For example, the words in the two sentences 'The cat sat on the man' and 'The man sat on the cat' are identical, yet the meaning of the sentences differs as a result of scrambling the order. In the same way, we need to exercise care when composing a mathematical sentence because we otherwise 'misread' it, and obtain an incorrect answer.

■ **Worked Example 3.1**

A man walks into a shop and asks for a bottle of milk and two packets of crisps. Each bottle of milk costs £1 and each packet of crisps costs 40p. How much does the man need to pay?

Methodology means the way we choose to perform a calculation or procedure.

We could just add up this bill in our heads and say £1.80, and we would be quite correct. But a formal **methodology** would say:

$$\text{total} = (\text{price per bottle of milk} \times \text{number of bottles})$$
$$+ (\text{price per packet of crisps} \times \text{number of packets})$$

The methodology for this simple example is to perform two multiplication calculations:

- First (price per bottle of milk × number of bottles).
- Next (price per packet of crisps × number of packets).

We obtain the final answer by adding up the two sums:

$$\text{total} = (£1 \times 1) + (40p \times 2) = £1.80 \quad ■$$

This example may be simple, but it illustrates the way that most calculations are made up of 'building blocks', each of which is itself a simple calculation.

To answer the sum 4×3, we could say either 'four times three' or 'three times four'. The answer is the same either way, if we count properly. But in any sum involving *two* operators, one will have priority over the other. If we ignore the priorities, we may obtain the wrong answer. For example, notice how we performed the multiplication steps first in Worked example 3.1 and only then did we add up the individual components. We say that multiplication, as an operator, has **priority** over addition.

In the language of mathematics, we say a calculation is **associative** if the answer does not depend on the order in which we do it. For example, the calculation $4 \times 3 \times 2$ is associative, because we obtain the same answer of 24 if we first multiply 4×3, then multiply its answer of 12 by 2, or if we multiply 4×2 by 3 or 3×2 by 4. We could write this, saying:

$$(4 \times 3) \times 2 = 4 \times (3 \times 2) = 3 \times (4 \times 2)$$

In a similar way, addition is associative: $3 + 7 + 1 = 7 + 3 + 1$.

Sums involving only addition or only multiplication are associative, but most sums are not. In fact, the majority of operators do not operate in a associative way. Accordingly, we need to formulate simple rules to ensure we perform a sum in the correct order.

> We say a calculation is associative if the answer does not depend on the order in which we do it.

■ Worked Example 3.2

Show that the problem $3 + 4 \times 5$ is not associative.

Firstly, we consider the case in which we perform the ADDITION step first: $3 + 4$. The result of this problem is then multiplied by 5:

$$3 + 4 = 7$$

then:

$$7 \times 5 = 35$$

Secondly, we consider the case in which we perform the MULTIPLICATION step first: 4×5. To the result of this problem, we then add 3:

$$4 \times 5 = 20$$

then:

$$20 + 3 = 23$$

The results of these two approaches to the same problem are clearly very different. ■

Worked example 3.2 helps illustrate the need for a series of rules when performing even simple problems. We know some of these rules already. For example, look at the use of the minus sign, which is a more complicated operator than $+$. The complexity arises because the operation it describes depends on the positions of the numbers or symbols: if we see the sum $4 - 3$, we automatically subtract the number positioned on the *right* of the operator from the number on the *left*. Swapping the 3 and 4 yields the wrong answer.

And we know from Worked example 3.2 that multiplication has priority over addition. When assigning priorities, we often start with the acronym **BODMAS**. In priority order, the letters stand for BRACKETS, OF, DIVISION, MULTIPLICATION, ADDITION, SUBTRACTION. The word 'OF' here is not a normal grammatical preposition, but implies more complicated operators, such as powers, roots, and those we meet in subsequent chapters. For example, we might say, 'The square *of* . . .' a number.

Before we start a calculation, write 'BODMAS' in large letters across the page:

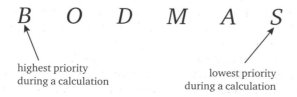

B   O   D   M   A   S

highest priority
during a calculation

lowest priority
during a calculation

■ Worked Example 3.3

A chemist has to perform a thermodynamic calculation, using the equation:

$$\Delta G^{\ominus} = \Delta H^{\ominus} - T \times \Delta S^{\ominus}$$

where $\Delta G^{\ominus}$ is the change in the Gibbs function during a reaction, $\Delta H^{\ominus}$ is the change in enthalpy during the same reaction, $\Delta S^{\ominus}$ is the entropy change, and T the temperature at which the reaction is performed.

Consider the Haber process to produce ammonia:

$$N_{2(g)} + 3H_{2(g)} \rightleftharpoons 2NH_{3(g)}$$

At 300 K, the relevant thermodynamic data are $\Delta H^{\ominus} = -92\,\text{kJ mol}^{-1}$ and $\Delta S^{\ominus} = 200\,\text{J K}^{-1}\,\text{mol}^{-1}$. What is the value of $\Delta G^{\ominus}$?

The equation above contains two operators, both '+' and '×'. Before we perform the calculation, we must first decide whether to perform the addition first, or the multiplication first. We will obtain the wrong answer for $\Delta G^{\ominus}$ if we use the operators in the wrong order.

Which operator has priority? When we look at the line, saying 'BODMAS', we see that the left-hand side is marked as highest priority and the right-hand end has a lower priority. 'M' for 'multiplication' is nearer to the left than is 'A' for 'addition'.

We therefore perform the MULTIPLICATION step *first*.

Inserting values into the equation:

$$\Delta G^{\ominus} = (-92\ \text{kJ mol}^{-1}) + (300\ \text{K} \times \Delta S^{\ominus} = 200\ \text{J K}^{-1}\ \text{mol}^{-1})$$

so:

$$\Delta G^{\ominus} = (-92\ 000\ \text{J mol}^{-1}) + (60\ 000\ \text{J mol}^{-1})$$

and:

$$\Delta G^{\ominus} = -32\ 000\ \text{J mol}^{-1}$$

As chemists, we might like to write this answer as $-32\,\text{kJ mol}^{-1}$. ■

■ Worked Example 3.4

Because an ion is charged, it will accelerate when it enters a magnetic field. Its acceleration is a. Its initial velocity is u. The distance travelled by the particle is s. The variables s, u, and a are related according to Newton's law of motion, where t is the time after entering the field:

$$s = ut + \tfrac{1}{2} at^2$$

In what order should we perform the calculation if we are to calculate a value of s?

Before we start, we notice how u and t are written together, but without an operator. The convention says we must assume variables are MULTIPLIED if written together without an apparent operator. We see how 'ut' is really $u \times t$ and '$\tfrac{1}{2}\,at^2$' is really $\tfrac{1}{2} \times a \times t^2$.

We now analyse the equation to see which operators it contains, and find ADDITION, MULTIPLICATION (twice), and a SQUARE function. In sequence:

1. The operator with the highest priority is the square function (within BODMAS, we call it OF, because 't^2' is a function OF t), so we would first calculate t^2.

The word data is plural. The singular is *datum* and never *datas*.

From now on, each time we meet one of these operators, we will print it in SMALL CAPITALS.

Remember:
A 'kilojoule' kJ is shorthand for 1000 J; see Chapter 1.

2. The operator with the next highest priority is MULTIPLICATION, so we perform the two MULTIPLICATION steps: $u \times t$ (i.e. ut) and $\frac{1}{2} \times a \times t^2$ (i.e. $\frac{1}{2} at^2$).

3. The last operator is ADDITION, so we add together the compound terms 'ut' and '$\frac{1}{2} at^2$' and obtain $s = ut + \frac{1}{2} at^2$. ∎

■ **Worked Example 3.5**

We can express the heat capacity C of a substance in terms of a power series:

$$C = a + bT + cT^2$$

where T is the temperature, and a, b, and c are constants obtained experimentally. In what order should we calculate C?

Equations of this type are sometimes called **virial equations**. As before, we first look to see which functions are involved. We see ADDITION, MULTIPLICATION (because bT means '$b \times T$' and cT^2 means '$c \times T^2$'), and a function OF, because T^2 is a function OF T.

1. In the BODMAS scheme, the operator with the highest priority is OF, so we first calculate T^2.

2. The operator with the next highest priority is MULTIPLICATION, so we calculate $b \times T$ and $c \times T^2$.

3. The final operator is ADDITION. We finally add together the individual terms, as $C = a + bT + cT^2$. ∎

■ **Worked Example 3.6**

We sometimes find it helpful to think of a chemical bond as behaving like a spring (see Fig. 3.1). The equation below describes the frequency at which the bond vibrates f, and relates the force constant k with the so-called 'reduced mass' of the vibrating species μ and the speed of light c:

$$f = \frac{1}{(2\pi c)} \sqrt{\frac{k}{\mu}}$$

If we know c, k, and μ, in what order should we perform the calculation of f?

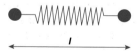

Figure 3.1 The vibration of a chemical bond has many analogies with two solid objects separated by an elastic spring. The length l extends from the centre of mass of one object to the centre of mass of the other.

Before we start this example, we notice the fraction immediately after the equals sign. We call fractions of the form '1 ÷ anything' a **reciprocal**.

We now look at the contents of this bracket. The bottom line clearly contains three terms, yet has no operators. It is most *unlikely* that $2\pi c$ is a single variable. Again, we are to read $2\pi c$ as though it was $2 \times \pi \times c$. We've discovered yet another shorthand notation. In fact, we could have written the equation as:

$$f = \frac{1}{(2 \times \pi \times c)} \times \sqrt{\frac{k}{\mu}}$$

We can now start planning the calculation.

1. From the laws of BODMAS, we give the highest priority to BRACKETS, because 'B' for BRACKET lies nearest the high-priority end. We must first calculate $(2 \times \pi \times c)$, because it is bracketed.

2. The next operator after BRACKETS is OF, which means 'a function of'. The simplest functions are powers and roots, so we next calculate a value of the term $\sqrt{k/\mu}$.

3. Next comes DIVISION, so we divide 1 by the bracket $(2\pi c)$.

4. Finally, the last task is to MULTIPLY together the two terms $1/(2\pi c)$ and $\sqrt{k/\mu}$. ■

■ **Worked Example 3.7**

The ideal-gas equation $pV = nRT$ is rarely realistic because of strong interparticle forces. The van der Waals equation is one of the better ways to take account of such forces:

$$\left(p + \frac{n^2}{V^2}a\right)(V - nb) = nRT$$

where the constants a and b are called the van der Waals constants, the values of which depend on the gas; n is the amount of gas.

To calculate a value of the compound variable 'nRT', what would be the correct order in which to calculate the left-hand side of the equation?

The a term in this equation reflects the strength of the interaction between gas particles, and the b term reflects the size of a particle.

There are six different operators in the left-hand side of the equation: we find ADDITION, SUBTRACTION, MULTIPLICATION, and DIVISION. And we also find BRACKETS. Before we start, we need to mention that the two terms V^2 and n^2 are best treated as functions OF V and n, each of which we consider within the BODMAS scheme under 'OF'.

1. Within the BODMAS scheme, the BRACKETS have the highest priority, so we calculate their contents first. The right-hand bracket is easy: we first multiply n with b (MULTIPLICATION) and only then SUBTRACT this compound term from V (i.e. saying '$V - nb$').

2. We next consider the left-hand bracket. We notice within this bracket both V^2 and n^2. These squares are functions OF V and n, so we square both terms before any further calculations.

3. Staying with the left-hand bracket, we next consider DIVISION, dividing the n^2 term by V^2, to yield n^2/V^2.

4. Next, again staying with this bracket, we consider MULTIPLICATION: we multiply a with n^2/V^2 to yield an^2/V^2, which we generally choose to write without the MULTIPLICATION sign as an^2/V^2.

5. To complete the left-hand bracket, we ADD the p term to n^2/V^2a, which completes the equation's left-hand side.

6. We finally MULTIPLY the 2 brackets together. ■

■ **Worked Example 3.8**

We now return to Worked example 3.5. It is sometimes appropriate to model the heat capacity C_p of a substance in terms of a **power series** (also called a *virial equation*). For chloroform (**I**), the appropriate series is

$$C_p = 91.47 + 7.5 \times 10^{-2}\, T - 10^{-6}\, T^2$$

where T is the thermodynamic temperature, and the factors relate to values of C_p expressed in units of J K^{-1} mol^{-1}. What is the value of C_p at 400 K?

The equation may confuse us because it contains so much detail. Accordingly, we will clarify its appearance by rewriting it slightly, with all the operators written in **bold** print:

$$C_p = (91.47) + (7.5 \times 10^{-2}) \times T - (10^{-6}) \times T^2$$

1. The operator of highest priority is the square (i.e. it's a function OF T). Our first step is to calculate T^2, which equals $400 \times 400 = 160\,000$. The equation becomes:

$$C_p = (91.47) + (7.5 \times 10^{-2}) \times T - (10^{-6}) \times 160\,000$$

2. Next in priority are the MULTIPLICATION steps. There are three. In order to emphasize the portions to be MULTIPLIED, we draw square brackets around the relevant portions:

$$C_p = (91.47) + [(7.5 \times 10^{-2}) \times 400] - [(10^{-6}) \times 160\,000]$$

which becomes

$$C_p = 91.47 + 30 - 0.16$$

3. Finally, we calculate C_p by ADDING the resultant terms, to yield:

$$C_p = 121.31 \text{ J K}^{-1} \text{ mol}^{-1} \ \blacksquare$$

Self-test 1

In each of the following, calculate the value of x using the BODMAS rule:

1.1 $\quad x = 3 \times 2 \times 6$

1.2 $\quad x = 6 \times 7 - 8 \times 2$

1.3 $\quad x = (2 + 3) \times 5$

1.4 $\quad x = 4 + 5^2 - 3^4$

1.5 $\quad x = \dfrac{4 \times 6}{7 \times 2}$

1.6 $\quad x = \dfrac{6 \times 9 - 2}{3 + 5 \times 2}$

1.7 $\quad x = 5 \times \sqrt{4 - 6 \times 2 + 44}$

1.8 $\quad x = \left(\dfrac{5 + 7}{6}\right)^2 - 56$

Additional problems

3.1 The molar mass of sulphuric acid is 100 g mol^{-1}, and the molar mass of water is 18 g mol^{-1}. What is the mass of 2 moles of sulphuric acid dissolved in 12 moles of water?

3.2 Calculate the molar mass of ferric nitrate nonahydrate, Fe(NO$_3$)$_3$.9(H$_2$O). The relevant atomic masses are Fe $= 56$, N $= 14$, O $= 16$, and H $= 1$ g mol^{-1}.

3.3 A chemist orders three bottles of chemicals: 12 of A and 7 of B. The mass of a single bottle of A is 500 g, while each bottle of B has a mass of 250 g. What is the overall mass of the bottles ordered?

3.4 The acceleration a of a proton inside a mass spectrometer may be calculated using the expression:

$$a = \frac{v - u}{t}$$

where the velocity changes from an initial value u to a different final value v during a time t. Calculate a if $u = 30 \, \mathrm{m\,s^{-1}}$, $v = 650 \, \mathrm{m\,s^{-1}}$, and $t = 3.9 \, \mathrm{s}$.

3.5 We define the gradient of a graph as:

$$\text{gradient} = \frac{y_2 - y_1}{x_2 - x_1}$$

where y values relate to vertical distances and x relates to horizontal distances. Calculate the gradient when $y_1 = 3.0$, $y_2 = 12.0$, $x_1 = 4.1$, and $x_2 = 5.5$.

3.6 The **entropy change** $\Delta S_{(cell)}$ for a cell depends on temperature, according to the following equation:

$$\Delta S_{(cell)} = nF\left(\frac{E_{(cell)2} - E_{(cell)1}}{T_2 - T_1}\right)$$

where n is the number of electrons transferred in the balanced redox reaction, F is the Faraday constant, $E_{(cell)}$ is the voltage of the cell, and T is temperature. Calculate a value of $\Delta S_{(cell)}$ if $n = 2$, $F = 98\,485 \, \mathrm{C\,mol^{-1}}$, $E_{(cell)}$ is 1.440 V at 298 K (i.e. at T_1) and 1.436 V at T_2 of 330 K.

3.7 A student sits three exams. The physical chemistry exam is worth 20 credits, and has a mark of 70%; the inorganic chemistry exam is worth 30 credits and has a mark of 63%; and the organic chemistry exam is again worth 30 credits and has a mark of 59%. The student therefore acquires 80 credits. What is the *overall* mark?

3.8 The **Kirchhoff equation** expresses the way an enthalpy change ΔH alters with temperature T:

$$\Delta H_2 = \Delta H_1 + C_p(T_2 - T_1)$$

where C_p is heat capacity. If $\Delta H_1 = 12 \, \mathrm{kJ\,mol^{-1}}$ at T_1 of 298 K, and $C_p = 31.2 \, \mathrm{J\,K^{-1}\,mol^{-1}}$, calculate the value of ΔH_2 when $T_2 = 330 \, \mathrm{K}$.

3.9 The **Einstein equation:**

$$E = mc^2$$

relates the energy released E when a mass m is converted entirely to energy, e.g. in a bomb; c is the speed of light. Calculate the energy released when 0.11 kg of uranium is converted. Take $c = 3.0 \times 10^8 \, \mathrm{m\,s^{-1}}$.

3.10 An **electrode potential** E varies with temperature T. For the silver bromide–silver redox couple, the value of $E_{AgBr,Ag}$ varies according to a virial expression:

$$E_{AgBr, Ag} = 0.07\,131 - 4.99 \times 10^{-4} \, T - 3.45 \times 10^{-6} \, T^2$$

Calculate the value of $E_{AgBr,Ag}$ when $T = 312 \, \mathrm{K}$.

4

Algebra III

Simplification and elementary rearrangements

Balancing the variables: why we need the equals sign, =

By the end of this chapter, you will know:

- That the equals sign '=' is not an operator, so it does not *do* anything to the variables.
- An equation is only complete when it contains an equals sign to relate the two sides of an equation.
- An equals sign tells us that the magnitudes of the two sides of the equation are identical in every way, but it tells us nothing about the content of either side.

It may seem trite to begin saying an equals sign '=' lies at the heart of our mathematics. The equals sign is like the verb at the heart of a sentence. For example, the two phrases 'the cat' and 'the mat' have nothing in common—we say they are unrelated—until we place some form of 'doing word' between them, something like 'the cat *sat* on the mat', 'the cat *ran away* from the mat', or 'the cat *liked* the mat'. The subject and the object are completely unrelated without the italicized verbs.

In just the same way, a variable such as 'pH' is not particularly interesting until we say something about it, and a number such as '3.2' is only a free-standing number until we attach it to something, using it to describe a variable we find interesting. But when we say 'pH = 3.2', suddenly we obtain information. Relating a variable (pH) to a numerical value (here, 3.2) allows us, as chemists, to know quite a lot.

The equals sign in an equation acts in just the same way as does the verb 'is'. In words, we could have said, 'the pH *is* 3.2', and we would have said neither more nor less than when writing, 'pH = 3.2'. They are the same.

The equals sign is not an operator. It does not do something to the variables in the way an operator such as + or × does. It merely gives us information.

Introducing the concept of simplification: collecting like terms

By the end of this chapter, you will know:

- Collecting together similar terms is one of the easiest ways of simplifying an equation.
- By 'similar terms' we mean algebraic portions that have the same form or content.

We collect terms together in everyday life. It simplifies otherwise difficult calculations. For example, we ask the grocer for a dozen eggs, rather than requesting them one egg at a time. We say 'six buses' drove past, rather than listing each one.

Collecting terms in this way saves us time and ink. It simplifies our lives. And we can simplify our maths in just the same way.

■ **Worked Example 4.1**

After analysing the composition of a reaction, the product contains the following atoms: C, C, C, C, C, H, H, H, H, H, C, C, O, O, H, and H. Simplify the way we express the composition by collecting together like terms.

The product contains three different elements: carbon, hydrogen, and oxygen. Adding up the C terms, we see the product contains seven carbons: call them 7C. It also contains two oxygens, 2O, and six hydrogens, 6H. We have simplified the maths by writing the composition as $7C + 6H + 2O$. In fact, the product is benzoic acid, (**I**).

I

This example may appear trivial, but adding terms together in this way enabled us to obtain empirical formulae from a **mass spectrum**. ■

....
The plural of **formula** is formulae (which we pronounce as *for-mew-lee*). The plural is never *formulas*.
....

■ **Worked Example 4.2**

A chemist obtains a mass spectrum, which reveals that the molecule responsible for the flavour of blackcurrants (**II**) contains the following fragments: CH_3, CH_3, CH_3, SH, CO, CH, CH_2, CH_2, CH_2, CH, C, and C. What is its empirical formula?

II

Adding together the terms, we have 11C, 18H, 1O, and 1S. We can write the formula as $C11 + H18 + 1O + 1S$, although most organic chemists prefer to write the formula of (**II**) in the form $C_{11}H_{18}OS$. ■

Self-test 1

Collect together the like terms, and thereby simplify the expressions:

1.1 $a + 2a + 3a + a$

1.2 $b + 4b + 6b + 22b$

1.3 $c + 2c + c + 4c + 4c + 4c + 7c$

1.4 $g + h + i + 5g + 6g + 2h + 6i$

1.5 $12g + 5g - 2g + 4g - 3g$

1.6 $6f - 5e + 6e - 7d + 4e + 3e + 8f$

1.7 $h + 9f - 3f - h + 5h - 6h + 4f$

1.8 $CH_3 + CH + SH + OH + CH_2 + CH_2$

Further simplifications: introducing the concept of balancing

By the end of this chapter, you will know:

- We say an equation is *balanced* when the collection of variables and numbers written to the left of the equals sign has the same value as the collection of variables and numbers written to its right.

- We can change the magnitude of either the variables or the numbers on the left-hand side of an equation provided we do *exactly* the same to the right-hand side.

- The equation is no longer balanced if we change only one side of the equation and retain the other. We say, *left-hand side ≠ right-hand side*.

Equations often *look* more difficult than they really are. It is wise therefore to make the equation look less intimidating. This chapter is designed to help us when we want to simplify an equation. The two simplest ways of simplifying an equation are factorizing (see Chapter 6) and cancelling, which is the subject of this chapter.

Care: Weight and mass are not identical. A mass represents an absolute gauge of how heavy an item is, while the weight depends on the extent of the Earth's gravitational pull.

Before we investigate any form of simplification, we need to learn a few 'ground rules'. Firstly, we need to see how the equals sign '=' lies at the very heart of an equation. An equation is not a real equation without one. We can think of the equals sign as acting much like an old-fashioned balance like that in Fig. 4.1. Just as the two arms of the balance need to be horizontal if the two pans contain the same mass, so the maths written to the left of the equals sign needs to balance the maths written to its right. If the two pans bear a different mass, the arms will tilt and we experience **inequality**. In a similar way, if we do something to one side of an equation, the two sides are not equal, meaning that we cannot write an equals sign between them.

Mathematically, we say:

$$mass_{\text{left-hand side}} = mass_{\text{right-hand side}}$$

if the two sides bear the same mass, and:

$$mass_{\text{left-hand side}} \neq mass_{\text{right-hand side}}$$

The symbol ≠ means 'is *not* equal to'.

if they differ.

The analogy of the arm balance is perfect. The two sides will always balance provided the two sides bear the same mass. The equality will continue to hold:

- Regardless of the actual magnitudes of the masses.
- Regardless of the identity of the materials on the balances.

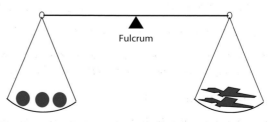

Fulcrum

Figure 4.1 An old-fashioned balance comprises two arms either side of a central **pivot** (or *fulcrum*). The left- and right-hand pans remain balanced only if they bear the same mass. Provided we do the same to both pans, the balanced remains equilibrated as demonstrated by the central arm being horizontal.

And we know the equality holds because the arm remains horizontal. In just the same way, an equals sign is a valid way of expressing the truth: 'the value of the left-hand side equals the value of the right-hand side'. This is true:

- Regardless of the actual magnitudes of the left- and right-hand sides.
- Regardless of the algebraic terms written on the equation's two sides.

The similarity extends. The arm of the balance remains horizontal, meaning the two pans bear the same mass, even if we change the contents of one pan, *provided we change the contents of the other pan by the same amount*. We can change the amounts on the two pans in one of two ways:

- We can add the same mass to both sides (we use the ADDITION operator).
- We can change the proportion on both sides (we use the MULTIPLICATION operator).

We can change the value of the right-hand side provided we perform the same operation to both sides. For example, we can ADD the same to both sides:

> ■ **Worked Example 4.3**
>
> The two arms of an old-fashioned balance are horizontal, so the contents of the two pans are equal. If we ADD a mass of 25 g to the left-hand side, what must we do to make it balance again?

To make the arms of the balance horizontal, we must do the same to both sides. As 25 g was ADDED to the left-hand side, we must also ADD 25 g to the right-hand side. ■

We can also SUBTRACT from both sides, e.g. to simplify an expression.

> ■ **Worked Example 4.4**
>
> After a reaction, the mass recorded was 10.4 g. Incorporated in that mass is the mass of the weighing boat, 0.3 g. What is the mass of product?

We start with the equation:

$$m_{\text{total}} = m_{\text{product}} + m_{\text{weighing boat}}$$

where m represents mass. The equation will continue to balance provided we do the same to both sides. We want to isolate the term m_{product}. Unfortunately, m_{product} does not appear on its own, but in conjunction with another mass, $m_{\text{weighing boat}}$. Since this additional term has been ADDED to the quantity we want, we perform the opposite operation, and SUBTRACT: we SUBTRACT the term $m_{\text{weighing boat}}$ from *both* sides:

$$m_{\text{total}} - m_{\text{weighing boat}} = m_{\text{product}} + (m_{\text{weighing boat}} - m_{\text{weighing boat}})$$

The equation is still balanced. The last two terms on the right-hand side (now in brackets) are the same. Any mathematical term minus itself is zero. We write:

$$m_{\text{total}} - m_{\text{weighing boat}} = m_{\text{product}} + (0)$$

There is no longer any need to write the last term in brackets. We see how the mass of the product is $m_{\text{total}} - m_{\text{weighing boat}}$, so:

$$m_{\text{product}} = 10.4 \, \text{g} - 0.3 \, \text{g} = 10.1 \, \text{g} \quad ■$$

Instead of saying 'opposite', we usually say we perform the **inverse** operation.

Self-test 2

Transform each of the following by ADDING or SUBTRACTING from *both* sides of the equation:

2.1 Rearrange $x + 9 = 12$, to obtain x on its own.

2.2 Rearrange $2x + 3x - 4x - 4v = 4v$, to obtain x on its own.

2.3 Rearrange $a = x - 12$ to make the term on the right, x.

2.4 Rearrange $a = x + 4c$ to make the term on the right, x.

2.5 Rearrange $4p = m_{one} - m_{two}$ to make the term on the right, m_{one}.

2.6 Rearrange $p = 4x - 3x + 5b$ to make the term on the right, x.

Simplification by cancelling

By the end of this chapter, you will know:

- Cancelling is a powerful way of simplifying equations.
- When we cancel, we look for like terms on the top and bottom of a fraction.
- If the like terms are multiples of each other, then they do not cancel completely, since a factor will remain.

The easiest way to simplify an equation is to **cancel**. We often find the first effect of rearranging an equation is to make it look more complicated.

Before we start, we will explore the statement 'A number divided by itself is 1.' A couple of examples readily demonstrate the statement's validity:

- One mole of sulphur contains one mole of sulphur.
- Four divides into four only once (in other words, '1 × 4' has a value of 4).

These statements are obviously true if somewhat trivial. We would not usually state these truths in such words, though; rather, we generally employ the symbols learnt from the previous chapter. We can state such relationships in a mathematical way, saying that a thing (any thing) divided by itself is **unity**, so '$x \div x = 1$'.

Anything divided by itself equals '1'.

■ Worked Example 4.5

Simplify the equation:

$$y = \frac{4cd}{2}$$

We start by rewriting the equation by separating the numbers from the algebraic letters:

$$y = \frac{4}{2} \times cd$$

In words, the fraction on the left is 'two into four'. It has a value of 2. We say the 4 and 2 **cancel**. We could have written:

$$y = \left(\frac{\cancel{2}}{\cancel{2}}\right) \times 2 \times cd$$

where the bracketed term '$2 \div 2$' has a value of 1 (where the diagonal lines are sometimes called **cancelling** lines). Accordingly we omit the bracket:

$$y = 2 \times cd$$

We normally omit the multiplication sign $\times$, and write just $2cd$. Alternatively, we can cancel without first splitting the number 4:

$$y = \frac{^{2}\cancel{4}}{\cancel{2}} \times cd$$

Here, note how the 4 on the top could be written as '2×2'. We cancel the 2 on the bottom with one of the 2s on the top (which leaves the second 2 untouched). When we collect together the remaining terms, we are left (as before) with $2 \times cd$. ∎

We need to note:

- By convention, we write the numbers *before* the letters: $2cd$ is acceptable, but neither of the representations $c2d$ or $cd2$ are standard.

- The cancelling lines are optional. Use them only if they help.

The additional 2 on the top row is the result of cancelling (i.e. $4 \div 2 = 2$).

■ **Worked Example 4.6**

A reaction proceeds in three steps:

$$A \xrightarrow{k_1} B \xrightarrow{k_2} C \xrightarrow{k_3} D$$

A is the initial reactant and D is the final product. B and C are reactive intermediates. Each of the three processes proceeds at a characteristic speed, so each is defined by is respective rate constant, k. Having performed a kinetics analysis, the equation describing the overall rate of reaction is:

$$\text{rate} = \frac{3k_1 k_2^2 k_3}{6k_1 k_2}$$

Simplify this equation by cancelling. ∎

We need to note:

- Each k term is a rate constant.
- Sometimes, we need to distinguish between similar-looking terms. One of the simplest ways is to employ subscripts, which we write to the right of the symbol.
- The subscripted numbers are only included to help us distinguish between the three different rate processes.
- k_2 means k of the second process. If it meant '2 times k', we would have written the number *before* the k and would have written the 2 on the *same* level as the k, rather than subscripted.
- The term k_2^2 means that the rate constant k_2 has been squared, i.e. $k_2^2 = k_2 \times k_2$.

Before we start, we will rewrite the expression. We first include *all* the multiplication signs, then collect the numbers together. To ease the analysis, we have grouped the like terms, one above another:

$$\text{rate} = \frac{3}{6} \times \frac{\cancel{k_1} \times k_2 \times k_2 \times k_3}{\cancel{k_1} \times k_2}$$

In just the same way as we cancelled numbers in the previous example, we can cancel the 3 and the 6 here. We will first cancel the '$k_1 \div k_1$' term, writing in its place '1':

$$\text{rate} = \frac{3}{6} \times 1 \times \frac{\cancel{k_2} \times k_2 \times k_3}{\cancel{k_2}}$$

We can similarly cancel the '$k_2 \div k_2$' term:

$$\text{rate} = \frac{3}{6} \times 1 \times 1 \times k_2 \times k_3$$

The 1×1 in the middle is a bit silly, so we'll ignore it from now on. Finally, we wish to simplify the fraction of $3 \div 6$. To this end, we note how the 6 on the bottom is the same as 3×2, which allows us to rewrite the fraction, using a different notation. We have:

$$\text{rate} = \left(\frac{\cancel{3}}{\cancel{3} \times 2} \right) \times k_2 \times k_3$$

The 3 on the top and bottom will cancel, leaving a factor of $\frac{1}{2}$. There is now no need to retain the brackets, so we obtain:

$$\text{rate} = \frac{1}{2} \times k_2 \times k_3$$

Accordingly, our expression collapses down to a manageable size. Omitting both multiplication signs, we write:

$$\text{rate} = \frac{1}{2} \, k_2 \, k_3$$

Self-test 3

By use of cancelling, simplify the following expressions:

3.1 $y = \dfrac{18c}{6}$

3.2 $y = \dfrac{2d}{4}$

3.3 $y = \dfrac{d}{4d}$

3.4 $y = \dfrac{bcd}{cde}$

3.5 $y = \dfrac{4}{3d} \times \dfrac{6}{20}$

3.6 $y = \dfrac{6a^2b}{6a}$

Introducing rearrangements: reversing the MULTIPLICATION and DIVISION operators

By the end of this chapter, you will know:

- As well as MULTIPLYING both sides of an equation, we can also DIVIDE.
- Before rearranging, we look for the operator. We then perform the reverse operation.

■ **Worked Example 4.7**

We wish to scale up a reaction involving thionyl chloride (**III**) with water:

$$\underset{\text{III}}{Cl{-}\overset{\displaystyle O}{\underset{\displaystyle O}{S}}{-}Cl} \;+\; 2\,H_2O \longrightarrow HO{-}\overset{\displaystyle O}{\underset{\displaystyle O}{S}}{-}OH \;+\; 2\,HCl$$

Before scaling up, we reacted 2 moles of water with 1 mole of (**III**). We want to react 30 moles of water. How much (**III**) do we need?

We start by writing an equation. We say:

$$1 \times n_{(III)} = 2 \times n_{(H_2O)}$$

where n is the amount of material involved in the reaction. In the second experiment, we clearly want *15* times as much water if we are to react 30 mol. We therefore multiply both sides of the equation by 15. We write:

$$15 \times 1 \times n(III) = 15 \times 2 \times n_{(H_2O)}$$

so:

$$15 \times n_{(III)} = 30 \times n_{(H_2O)} \quad \blacksquare$$

> It is usual practice to omit the multiplication signs, and write $15\, n_{(I)} = 30\, n_{(H_2O)}$.

■ **Worked Example 4.8**

We wish to scale up a reaction. In our first attempt, we made 1 mol of nicotine (**IV**), which had a mass of 165 g. We now want to make 4.2 mol of (**IV**). How much nicotine will we make?

IV

We start by writing the equality:

$$1\,mol\,of\,(IV) = 165\,g$$

We can MULTIPLY either side of the equation by any **factor** we like, provided we MULTIPLY *both* sides by the same factor. In this case, we want 4.2 moles of (**IV**), so here we MULTIPLY both sides by 4.2. We write:

$$4.2 \times 1\,mol\,of(IV) = 4.2 \times 165\,g$$

So the mass of 4.2 mol of (**IV**) will be:

$$4.2 \times 165\,g = 693\,g \quad \blacksquare$$

> A **factor** is either one of the components within a number or algebraic expression, or a term having a constant value by which we multiply a variable.

Treating the equals sign as a type of balance is so powerful a tool that we need not restrict ourselves to MULTIPLYING both sides: we can DIVIDE both sides.

■ **Worked Example 4.9**

We wish to make some ethyl vanillin (3-ethoxy-hydroxybenzaldehyde, **V**) to give a vanilla flavour to some food. We wish to scale down the reaction: we started by making 1 mol of (**V**) and made 166 g, but now we want to decrease the amount 30-fold. How much (**V**) do we make?

We start with the equation:

$$1\,mol = 166\,g$$

We then DIVIDE both sides by 30:

$$\frac{1\,mol}{30} = \frac{166\,g}{30}$$

> Ethyl vanillin is added to food to give a strong flavour of vanilla.

so 1/30th mol of (**V**) has a mass of 5.53 g.

V

We see how treating the equals sign in an equation as a balance is a powerful way of scaling up or down, as achieved by MULTIPLYING or DIVIDING. ∎

■ **Worked Example 4.10**

The ideal-gas equation:

$$pV = nRT$$

relates the pressure and temperature T of n moles of gas housed in a volume V. We call the term R the gas constant.

By use of cancelling, obtain an equation in which p is the **subject**.

Before we start, we need to define the new term above. We always write the **subject** in the form, 'Something on its own $= \ldots$'. The remainder of the equation is then called the **object**. This way of labelling an equation again reminds us of English grammar, and emphasizes how we can think of an equals sign as behaving a bit like the *verb*, 'to be'.

We want the symbol for pressure written on its own, in the form '$p = \ldots$'. To this end, we note how the symbol pressure p is presently MULTIPLIED by the symbol for volume V. We need to remove the V term to ensure p appears on its own, so we cancel the V term.

The operator performing the opposite function to MULTIPLY is DIVIDE. To remove the V term multiplied by the p term, we must divide the **compound variable** pV by V.

> Two variables multiplied together are called a **compound variable**.

Before we proceed further, we need to retain the equals sign ' $=$ '. We must therefore DIVIDE both sides of the ideal-gas equation by V. If we DIVIDED only the left-hand side by V, the two sides would no longer have the same value, so we DIVIDE both sides by V:

$$\frac{pV}{V} = \frac{nRT}{V}$$

Look at the left-hand side. It contains two V terms: one each on the top and the bottom. We can therefore rewrite the equation slightly:

$$p \times \left(\frac{\cancel{V}}{\cancel{V}} \right) = \frac{nRT}{V}$$

This version is identical to the one above except we have collected together the two V terms. Since 'V DIVIDED by V' is a number divided by itself, the bracket has a value of 1. We therefore rewrite the equation as:

$$p \times 1 = \frac{nRT}{V}$$

This MULTIPLYING by 1 is silly. Clearly, $1 \times p$ is the same as p on its own, so we write the left-hand side as 'p' on its own. Therefore, when p is the subject, we obtain:

$$p = \frac{nRT}{V} \ \blacksquare$$

Notice how, in this example, we started with V on the top left and ended with V positioned on the bottom right. This is a general result: when we start with a fraction, and rearrange by MULTIPLYING or DIVIDING in this way, the variable we move changes from one side of the equation to the other, and also moves from top to bottom or vice versa: look what happened when we MULTIPLY both sides by **b**:

$$\frac{a}{b} \neq \frac{c}{d} \quad \rightarrow \quad \frac{a}{1} = \frac{\boldsymbol{b} \times c}{d}$$

The direction of the arrow helps explain why we use the phrase **cross-multiply**.

Similarly, when we DIVIDE, the variable moves from one side of the equation to the other. Look what happened when we divide both sides by **c**:

$$\frac{a}{b} \neq \frac{c}{d} \quad \rightarrow \quad \frac{a}{b \times c} = \frac{1}{d}$$

> We need the 1 on the top of the right-hand fraction here, because a fraction must have *something* on top. The 1 here comes from rewriting the top line as $c \times 1$ before we divided.

■ Worked Example 4.11

n moles of solute are dissolved in a volume V of solvent to make a solution of concentration $[C]$. n, V, and $[C]$ are related by the equation:

$$[C] = \frac{n}{V}$$

By use of cancelling, obtain an equation in which n is the subject.

The symbol for the amount of substance n has been DIVIDED by the symbol for volume, V. To obtain the n term on its own, we must get rid of the V term. We reverse this operation, and multiply *both* sides of the equation by V:

$$[C] \times V = \frac{n \times V}{V}$$

As before, we see two V terms in the equation, both on the top and bottom of the right-hand side. We group them together:

$$[C] \times V = n \times \left(\frac{\cancel{V}}{\cancel{V}} \right)$$

The two V terms cancel, because anything DIVIDED by itself equals 1, so $V \div V = 1$:

$$[C] \times V = n \times 1$$

Omitting the multiplication sign, we obtain $n = [C]V$. ■

We need to note how in both these examples:

- We looked at the equation before starting.
- By inspection, we saw which mathematical OPERATIONS were involved, i.e. the process by which the variable of interest transmutes into the answer. In each case, it had either been MULTIPLIED or DIVIDED by something else.
- We then applied the *inverse* operator, in order to remove the other variables. In this way, we make the variable of interest the subject of the equation.
- Sometimes we cancel in order to obtain a number, rather than to rearrange an equation.

> We again follow the convention that tells us to write the subject on the *left* of the algebraic expression.

Self-test 4

Rearrange the following equations, in each case to make x the subject:

4.1 $y = mx$

4.2 $12x = 4$

4.3 $240x = 50y$

4.4 $abx = 12$

4.5 $mgh = 55x$

4.6 $mgx = 55h$

Collect together the following terms, then rearrange, again to make x the subject:

4.7 $5x + 6x = g$

4.8 $6x + 4x + x - 3x = 34$

4.9 $5f + 3f - 2f = 4x + 9x$

4.10 $p + q + 4q - 7q = 6x$

Rearrangements involving simple powers and roots

By the end of this chapter, you will know:

- We rearrange a power by taking the appropriate root.
- We arrange a root by taking the appropriate power.

When we have an operator such as a power (which is treated within BODMAS as a function 'OF'), we first identify the nature of the power, then perform the reverse operation. For example, we reverse a simple square by take the square root, reverse a cube by taking the cube root, etc.

■ **Worked Example 4.12**

We devise a kinetics experiment to follow the dimerization of the fluorene radical anion (**VI**). The rate of dimerization follows the relationship:

rate $= k[(\text{VI})]^2$

where k is a so-called *rate constant*.

VI

Rearrange this equation to make the concentration [(**V**)] the subject.

We need to look carefully at the equation before rearranging it, to see what has happened to the concentration, [(**VI**)].

- It has first been SQUARED, to yield $[(\text{VI})]^2$. Within the BODMAS scheme, we must consider this squaring to be a function, OF.
- Next, $[(\text{V})]^2$ was MULTIPLIED by k.

To rearrange, we invert each step, and do so in the *reverse* order to which it was first performed. We therefore start by DIVIDING both sides by the rate constant k:

$$\frac{\text{rate}}{k} = \frac{k[(\text{VI})]^2}{k}$$

From now on, we will simplify the *presentation* and omit the cancelling steps. By simplifying, we must not forget that we have actually cancelled the k terms on the right-hand side of the equation.

Clearly, we cancel the two k terms on the right-hand side to yield:

$$\frac{\text{rate}}{k} = [(\text{VI})]^2$$

To obtain the concentration $[(\text{VI})]$ from its square $[(\text{VI})]^2$, we next perform the inverse function to SQUARING by taking the SQUARE ROOT of both sides:

$$\sqrt{\frac{\text{rate}}{k}} = \sqrt{[(\text{VI})]^2}$$

The square root of a square reverts back to the original function, just as $\sqrt{(4)^2} = 4$. Accordingly, we obtain:

$$[(\text{VI})] = \sqrt{\frac{\text{rate}}{k}}$$

If this last step may look a little advanced, try a few sums involving straightforward numbers. For example, the square of the square root of 16 is 16, i.e. $(\sqrt{16})^2$ because we know how $\sqrt{16}$ is 4, and 4^2 is 16. ∎

■ **Worked Example 4.13**

Sodium chloride crystallizes in a simple cubic structure. Each side of the cube—we call it a **unit cell**—has a length of l. If the unit cell's volume is $1.728 \times 10^{-30}\,\text{m}^3$, calculate the length of each side.

In this example, we first rewrite the data in the form of an equation. Because the volume is that of a cube, $l^3 = 1.728 \times 10^{-30}$. To reverse this operation, we identify the operator and perform the reverse operation: the operator is 'cube of', so we take the cube root. Clearly, we must take the cube root of both sides:

$$l = \sqrt[3]{1.728 \times 10^{-30}\,\text{m}^3}$$

so

$$l = 1.2 \times 10^{-10}\,\text{m} \quad ∎$$

> The small 3 to the left of the root sign means 'cube root'.

Self-test 5

Rearrange the following, in each case making x the subject:

5.1 $y = 3x^2$

5.2 $4y = ax^3$

5.3 $by = ax^4$

5.4 $y = \sqrt{x}$

5.5 $ay = 4\sqrt{x}$

5.6 $y = \sqrt[4]{x}$

5.7 When a rotating disc electrode (RDE) is immersed in solution, electrolysis only occurs within the thin layer of solution adjacent to the electrode's surface. We call this layer, the **Nernst depletion layer**, δ.

The thickness of δ is given by the equation:

$$\delta = \frac{1.61v^{1/6}D^{1/3}}{\sqrt{\omega}}$$

where D is the diffusion coefficient, v the viscosity, and ω is the rotation speed.

Rearrange the equation to make ω the subject.

> **Aside**
>
> To perform this cube root on a pocket calculator:
>
> 1. Type in '1.728×10^{-30}'.
> 2. Press the $x^{1/y}$ button.
> 3. Type in '3'.
> 4. Press '='.

Additional problems

The answers to some of these questions will be obvious. Nevertheless, determine the answer using the methodology outlined in this chapter.

4.1 The mass of 4 moles of aspirin is 1424 g. Calculate the mass of 1 mole.

4.2 The volume of 1 mole of ethanol is 45 cm^3. Calculate the volume occupied by 6.2 moles of ethanol.

4.3 The ideal-gas equation:

$$pV = nRT$$

relates the volume V of an ideal gas to the temperature T, the amount of gas n, and the pressure p. R is the gas constant. Rearrange this equation to make T the subject.

4.4 The Beer–Lambert law:

$$A = \varepsilon[C]l$$

relates the optical absorbance A of a coloured sample to its concentration $[C]$, the thickness of the sample l, and the molar absorptivity ε. Rearrange the equation to make $[C]$ the subject.

4.5 The equation relating temperatures in kelvin T and Celsius t is:

$$T = t + 273.15$$

Rearrange this expression to make t the subject.

4.6 A chemist weighs some compound onto a balance, and records an overall mass of 12.443 g. If the weighing boat has a mass of 0.250 g, what is the mass of sample?

4.7 Consider the simple second-order esterification reaction: acid + alcohol $\rightarrow$ ester + water. The rate equation for the reaction is:

$$\text{rate} = k_2[\text{acid}][\text{alcohol}]$$

Rearrange this rate equation to make the second-order rate constant k_2 the subject.

4.8 The area A of a square electrode is 7.2 cm^2. Calculate the length of each side of the square l (remember $A = l^2$).

4.9 The equation relating the changes in the Gibbs function $\Delta G^{\ominus}$, enthalpy $\Delta H^{\ominus}$, and entropy $\Delta S^{\ominus}$ is:

$$\Delta G^{\ominus} = \Delta H^{\ominus} - T\Delta S^{\ominus}$$

where T is the temperature. Rearrange the equation to make $\Delta H^{\ominus}$ the subject.

4.10 The Clausius equality relates minuscule changes in energy dq with temperature T and the change in entropy ΔS:

$$\Delta S = \frac{dq}{T}$$

Rearrange this expression to make dq the subject.

5

Algebra IV

Rearranging equations according to the rules of algebra

Rearranging equations of more than one operator: BODMAS

By the end of this chapter, you will know:

- If we have an equation of the type '*x* = . . .', we say *x* is the subject of the equation.
- When we need to rearrange complicated equations, we firstly identify the operators present.
- Secondly, we assign priorities according to the BODMAS scheme.
- Thirdly, starting with the operator of lowest priority, we perform the inverse operation to *both* sides of the equation.
- We invert each operation in turn until the rearrangement is complete.

The great theoretical physicist Albert Einstein once said, 'Reading a mathematical equation is much like reading a detective novel: the symbols are the clues and we are the detectives who use them to solve the mystery.' We will learn in this chapter what are the clues, and how to read them. The 'mystery' is what happened to a variable en route to becoming the final equation.

We can often perform a rearrangement in our head without needing the methods outlined below, i.e. without formal logic. Many examples here fall into that category. But it is *always* better to include all the steps in the algebra and get the right answer rather than leave them out, and get it wrong. The simple examples here are included to show how formal logic yields the same answer as the one we would have reached without formal logic. We will also look at examples in which intuition is wholly inadequate.

The examples in previous chapters were all straightforward since each contained a single operator. Each time we rearranged an equation, we first identified the operator, and then performed its inverse. We will use a similar approach here. Unfortunately, however, we cannot do that straightaway because these equations are more complicated in so far as they each contain more than one operator.

Chapter 2 introduced us to the BODMAS rule. There, we learnt the correct order in which to perform a series of operations. For example, we could only use SUBTRACTION and ADDITION operators *after* we had MULTIPLIED or DIVIDED. In the same way, when we rearrange an equation involving different operators, we must learn the correct step-by-step sequence in which the rearrangement proceeds.

In this chapter, we are not performing a series of operations to obtain an answer; rather, we wish to **rearrange** the equation. To do so, we will employ the methodology introduced in the previous chapter: firstly, we look at the equation to see what had happened to the variable of interest. The equations in this chapter have a minimum of two operators.

Secondly, we will rank operators into a hierarchy. It is our job therefore to unpick the equation, bit by bit. The order of the operators in the BODMAS scheme is the same whether we perform a calculation or rearrangement. The sole difference is that we work *backwards* during rearrangement.

> It is *always* better to include all the steps and get the right answer, rather than leave them out, and get it wrong.

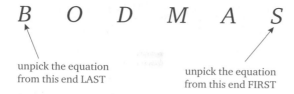

unpick the equation
from this end LAST

unpick the equation
from this end FIRST

■ **Worked Example 5.1**

The strong magnets inside a mass spectrometer accelerate a charged ion. After a time of t, the ion's velocity has increased from u to v. The acceleration is a. The variables are interrelated according to the equation:

$$v = u + at$$

Rearrange the equation to yield an expression for the acceleration a, expressed in terms of the other variables, i.e. obtain a final equation of the form $a = \ldots$.

Strategy:

(i)   We first look at the equation to ascertain which operators are involved.

(ii) We then decide which operators have priority, according to the BODMAS scheme.

(iii) We ascertain the order in which the equation will have been compiled, i.e. if we started with a, what was the sequence of steps which occurred in order to obtain the final equation?

(iv) We reverse each operation, starting with the operator of *lowest* priority.

(i)   There are two operators:
      • MULTIPLICATION (yielding $a \times t$, which we write as at).
      • ADDITION (which combines the two terms u and at to form $u + at$).

(ii) Within the BODMAS scheme, the MULTIPLICATION operator has the higher priority, so ADDITION has a lower priority.

(iii) The value of a was MULTIPLIED by t to form the compound variable at. To this compound variable, u was ADDED to yield the final velocity v.

(iv) We start to 'unpick' the equation in the order dictated by the BODMAS scheme. Since ADDITION is the lowest priority, we first remove the term ADDED to at by performing the inverse operation. Since u was ADDED to at, we SUBTRACT u from both sides the equation:

$$v - u = \boxed{u - u} + at$$

Value is equal to zero

> The equation would not have balanced if we had not performed the same operation to both sides.

In writing a term 'u' on both sides, we demonstrate that we remembered how the equals sign '$=$' at the centre of the equation requires that we do the same to both sides. The value of $u - u$ on the right-hand side is clearly 0 (as we intended), and effectively removes u as a variable from that side of the equation.

The equation becomes:

$$v - u = at$$

To obtain a on its own, we next apply the MULTIPLICATION step, which is the only operator remaining. The a term that we want has been MULTIPLIED by t. To obtain a on its own, we must perform the inverse (i.e. opposite) operation to

MULTIPLICATION, so we DIVIDE by t. The equals sign '$=$' tells us to do the same to both sides of the equation:

$$\frac{v - u}{t} = \frac{at}{t}$$

We must remove the two t terms on the right-hand side. To do so, we could simply say $t \div t = 1$. More formally, we rewrite the equation in a slightly different way:

$$\frac{v - u}{t} = a \times \left(\frac{t}{t} \right)$$

This equation is identical to that above except we have reformatted its right-hand side. The bracketed term has a value of 1. In effect, it says 't goes into t once'. The bracket may be substituted with 1. We say their values **cancel**, as discussed more fully in Chapter 3.

Having removed the two t terms, we obtain:

$$\frac{v - u}{t} = a \times 1$$

Any number or algebraic term divided by itself has a value of 1.

We say the acceleration a is the **subject** of this equation. It would be ludicrous to write '$\times 1$', so we'll not mention the factor of 1 any further.

We might choose to write this answer with a bracket, as:

$$a = \left(\frac{v - u}{t} \right) \ \blacksquare$$

Notice how time is written as a lower-case letter t and temperature is written as an upper-case letter T. Both are written in italic type because they are variables.

Adding brackets causes no material difference to the equation, but some people think it looks neater.

We need to note:

- When we rearrange in order to make one term the subject, the usual practice is to write the final line with the subject on the *left*.

■ Worked Example 5.2

The value of the Gibbs function $\Delta G^{\ominus}$ of reaction varies quite strongly with temperature T, according to the **Gibbs–Helmholtz equation**:

$$\frac{\Delta G^{\ominus}}{T} = \frac{\Delta H^{\ominus}}{T} + c$$

By writing only $\Delta G^{\ominus}$ on the left-hand side of an equation, we make it the **subject** of the equation.

where $\Delta H^{\ominus}$ is the change in enthalpy of the same reaction and c is a constant.
Write an expression in which $\Delta G^{\ominus}$ is the only term on the left-hand side.

We want to make $\Delta G^{\ominus}$ the subject of the equation. Before attempting an answer, we analyse the equation to see how the value of ΔG on the left-hand side has been changed. In this example, $\Delta G^{\ominus}$ has been DIVIDED by the temperature T. We therefore rearrange the equation in such a way that the T term disappears from the left-hand side. The opposite function to DIVIDE is MULTIPLY, so we MULTIPLY both sides of the equation by T:

$$T \times \frac{\Delta G^{\ominus}}{T} = T \left(\frac{\Delta H^{\ominus}}{T} + c \right)$$

We place the brackets round the right-hand term '$\Delta H^{\ominus} \div T + c$' because, at this stage, we must treat it as an indivisible whole. It is a **compound variable**. Accordingly, when we MULTIPLY both sides of the equation by temperature T, we MULTIPLY the *whole* of the right-hand side by T. The brackets help avoid any potential confusion.

We now return to the left-hand side. We can rewrite the equation slightly:

$$\left(\frac{T}{T}\right) \times \Delta G^{\ominus} = T\left(\frac{\Delta H^{\ominus}}{T} + c\right)$$

The term in brackets on the left-hand side can be equated to 1 because any number divided by itself is unity. The resultant expression for $\Delta G^{\ominus}$ therefore becomes:

$$\Delta G^{\ominus} = T\left(\frac{\Delta H^{\ominus}}{T} + c\right)$$

We give the name **cancelling** to the process of manipulating an equation to ensure that one component term takes a form like $T \div T$. ■

> We could multiply out the right-hand side, and write '$\Delta H^{\ominus} + Tc$'. The expression here is just as valid, although the version with brackets looks a little neater.

■ Worked Example 5.3

The Hückel rule says that a fused-ring system is aromatic if the number N of its π-electrons obeys the equation:

$$N = 4n + 2$$

where n is an **integer** such as 1, 2, 3, etc. For example, $n = 4$ for tetracene (**I**). Rearrange the equation to make n the subject.

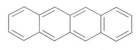

I

> **Integer** is the mathematical word meaning 'whole number'.

Two operators are apparent, both ADDITION and MULTIPLICATION. The BODMAS rule says that ADDITION has the lower priority, so we reverse the ADDITION step, and perform the inverse operation SUBTRACTION to both sides:

$$N - 2 = 4n(+2 - 2)$$

The bracket on the right-hand side disappears, as intended. We obtain:

$$N - 2 = 4n$$

One operator remains on the right-hand, since n has been MULTIPLIED by 4. We therefore perform the inverse operation of DIVIDE. We DIVIDE both sides by 4:

$$\frac{N - 2}{4} = \frac{4n}{4}$$

The two factors of 4 on the right-hand side cancel, leaving n as the subject of the equation:

$$n = \frac{N - 2}{4} \quad ■$$

Many people are tempted to cancel the 2 and 4 on the left-hand side. In fact we cannot cancel them, because the 2 on the top is not a multiplication factor. It has merely been SUBTRACTED from N.

■ Worked Example 5.4

Consider the fraction $n = (10 - 2)/(8 - 4)$. Show how inappropriate cancelling yields an incorrect result.

Firstly, we evaluate the simple sums on top and bottom:

$$n = \frac{10 - 2}{8 - 4} = \frac{8}{4}$$

Clearly, then, $n = 2$.

Secondly, we cancel inappropriately. The most common way of doing this is to note how 2 goes into 4 twice, and say:

$$n = \frac{10 - \cancel{2}}{8 - \cancel{4}_2}$$

so

$$n = \frac{10}{8 - 2} = \frac{10}{6}$$

This time, n has a value of 1.66, which differs greatly from the correct answer above. ∎

Sometimes we want an equation in which the subject is not a single term but a phrase or compound variable, see Worked example 5.2.

■ **Worked Example 5.5**

The **Kirchhoff equation** tells us how the enthalpy change accompanying a reaction ΔH varies according to the temperature T at which we perform it. If the enthalpy change at a temperature T_1 is ΔH_1 and the enthalpy change at a temperature T_2 is ΔH_2, then:

$$\Delta H_2 = \Delta H_1 + C_p(T_2 - T_1)$$

where C_p is the heat capacity at constant pressure.

Rearrange the Kirchhoff equation to make C_p the subject.

Before we start, we look critically at the equation to discern what operators influence C_p. We see it has been MULTIPLIED by the bracket $(T_2 - T_1)$ to form '$C_p \times (T_2 - T_1)$'. ΔH_1 was then ADDED to this resultant compound variable.

> We ignore the two operators SUBTRACTION and BRACKETS because the subtraction only occurs within the bracket, and we move the bracket without affecting its contents in any way.

(i) When taking apart this equation, the operator of lowest priority is ADDITION (so ΔH_1 has been ADDED), so we perform the reverse operation of SUBTRACTION, and first SUBTRACT ΔH_1 from both sides to yield:

$$\Delta H_2 - \Delta H_1 = C_p \times (T_2 - T_1)$$

> In fact, before cancelling, the right-hand side of the equation reads:
>
> $$(\Delta H_1 - \Delta H_1) + C_p \times (T_2 - T_1)$$

(ii) Having removed the ΔH term, we see how C_p was MULTIPLIED by the bracket $(T_2 - T_1)$, so we now need to reverse the MULTIPLICATION step by performing the inverse operation of DIVISION. In this case, we DIVIDE both sides by the *whole of the bracket*:

$$\frac{\Delta H_2 - \Delta H_1}{(T_2 - T_1)} = C_p \frac{\cancel{(T_2 - T_1)}}{\cancel{(T_2 - T_1)}}$$

The twin brackets on the right-hand side cancel (as intended) to leave:

$$C_p = \frac{\Delta H_2 - \Delta H_1}{(T_2 - T_1)} \quad ∎$$

We need to note:

- The laws of algebra permit us to ADD, SUBTRACT, DIVIDE, or MULTIPLY by a complete bracket, provided we do not alter the bracket's contents in any way.
- We generally try not to divide by a bracket if the bracket contains a fraction.
- The brackets on the bottom of the last line are superfluous, and can be omitted.

We now consider what amounts to a special case, and learn how to proceed when rearranging equations that contain negative terms.

■ **Worked Example 5.6**

A reaction is thermodynamically feasible provided the value of the Gibbs function $\Delta G^{\ominus}$ is negative. The magnitude of $\Delta G^{\ominus}$ depends on the change in the enthalpy of reaction $\Delta H^{\ominus}$ and the entropy change $\Delta S^{\ominus}$, according to the equation:

$$\Delta G^{\ominus} = \Delta H^{\ominus} - T\Delta S^{\ominus}$$

where T is the thermodynamic temperature.

Rearrange this equation to make the change in entropy, $\Delta S^{\ominus}$, the subject.

As usual, before we start, we look to identify the operators present, and assess the order in which they are invoked. We see how $\Delta S^{\ominus}$ was first MULTIPLIED by the temperature T to form $T\Delta S^{\ominus}$, and, next, this compound variable $T\Delta S^{\ominus}$ was SUBTRACTED from $\Delta H^{\ominus}$.

As usual, we will invert the roles of these operators, and deal with the operator of lowest priority first. In fact, we can rearrange this equation in two ways: one long and the other shorter but a little more involved.

The longer (but simpler) route

We start by inverting the SUBTRACTION step by ADDING the compound variable $T\Delta S^{\ominus}$ to both sides:

$$\Delta G^{\ominus} + T\Delta S^{\ominus} = \Delta H^{\ominus}(-T\Delta S^{\ominus} + T\Delta S^{\ominus})$$

The bracketed portion on the right-hand side clearly equates to 0 and vanishes, which is what we wanted. We are left with the equation:

$$\Delta G^{\ominus} + T\Delta S^{\ominus} = \Delta H^{\ominus}$$

Next we SUBTRACT $\Delta G^{\ominus}$ from both sides in much the same way, to yield:

$$(\cancel{\Delta G^{\ominus}} - \cancel{\Delta G^{\ominus}}) + T\Delta S^{\ominus} = \Delta H^{\ominus} - \Delta G^{\ominus}$$

Again, the bracketed portion cancels, yielding:

$$T\Delta S^{\ominus} = \Delta H^{\ominus} - \Delta G^{\ominus}$$

Finally, to invert the MULTIPLICATION of $\Delta S^{\ominus}$ by T, by performing the opposite operation, we DIVIDE both sides by T:

$$\frac{T\Delta S^{\ominus}}{T} = \frac{\Delta H^{\ominus} - \Delta G^{\ominus}}{T}$$

The two T terms on the left-hand side cancel, to yield:

$$\Delta S^{\ominus} = \frac{\Delta H^{\ominus} - \Delta G^{\ominus}}{T}$$

The shorter (but more involved) route

In the shorter route, as with the longer route, we start by identifying the operators present. Again, we start by noting how $\Delta S^{\ominus}$ is MULTIPLIED by a function of temperature, but this time we say $T\Delta S^{\ominus}$ is MULTIPLIED by -1 to yield $-T\Delta S^{\ominus}$. $\Delta H^{\ominus}$ is then ADDED to the compound variable $T\Delta S^{\ominus}$.

To make ΔS the subject, we first subtract $\Delta H^{\ominus}$ as in the longer version, to obtain:

$$\Delta G^{\ominus} - \Delta H^{\ominus} = -1 \times T\Delta S^{\ominus}$$

Note how the factor of -1 persists. The right-hand side has been MULTIPLIED by $-1 \times T$, so we perform the opposite operation and DIVIDE both sides by $-1 \times T$:

$$\frac{\Delta G^{\ominus} - \Delta H^{\ominus}}{-1 \times T} = \frac{-T\Delta S^{\ominus}}{-1 \times T}$$

The $-1 \times T$ on the bottom of the fraction on the right-hand side is the same as $-T$; and the two $-T$ terms on the right-hand side soon cancel to yield $\Delta S^{\ominus}$, so:

$$\Delta S^{\ominus} = \frac{\Delta G^{\ominus} - \Delta H^{\ominus}}{-T}$$

This answer ought to be the same as that derived above using the longer route, because each derives from the same equation, $\Delta G^{\ominus} = \Delta H^{\ominus} - T\Delta S^{\ominus}$, and each expresses $\Delta S^{\ominus}$ as their subject. While they look slightly different, in fact they are the same: the only difference is that the right-hand sides have been rearranged somewhat.

To demonstrate the equality of the two equations, we will take the version immediately above, and multiply both top and bottom by -1. We write:

$$\Delta S^{\ominus} = \left(\frac{-1}{-1}\right) \times \frac{\Delta G^{\ominus} - \Delta H^{\ominus}}{-T}$$

Remember:
$-1 \times -1 = +1$, so
$-1 \times -T = +T$.

which is the same equation because the new bracketed term has a value of 1, and multiplying by 1 will not change the values. The value of -1×-1 on the bottom equals $+1$.

We then MULTIPLY the top line to yield:

$$\Delta S^{\ominus} = \frac{-\Delta G^{\ominus} + \Delta H^{\ominus}}{T}$$

Rearranging the top line yields:

$$\Delta S^{\ominus} = \frac{\Delta H^{\ominus} - \Delta G^{\ominus}}{T}$$

so the two versions of the equation are the same.

This type of rearrangement, multiplying by -1, is effectively just a dodge and represents an easy way to effect a rearrangement. It only works because $-1 \times -1 = +1$. ∎

Self-test 1

Rearrange the equations in 1.1–10 to make x the subject:

1.1 $y = 5x - 4x + 7x + 1$

1.2 $y = 3x - 7$

1.3 $y = 5 - 4x$

1.4 $y = \dfrac{x - 4}{7}$

1.5 $y = 8(x - 1)$

1.6 $y = 8x(b - 1)$

1.7 $y = d/x$

1.8 $y = \dfrac{3}{x - 2}$

1.9 $y = \dfrac{a - x}{4d}$

1.10 $y = \dfrac{x - 9}{c + d}$

1.11   A temperature can be expressed either in degrees Celsius (C) or in degrees Fahrenheit (F). These temperature notations are linked by the equation:

$$C = \frac{5}{9} \times F - 32$$

Rearrange to make F the subject.

» Continued . . .

⟫ *Self test 1 Continued...*

1.12   Consider the right-angled triangle below:

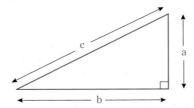

Pythagoras's theorem says the lengths of the sides follow the relationship $c^2 = a^2 + b^2$.

Rearrange Pythagoras's equation to make b the subject.

Rearrangements involving simple powers and roots

By the end of this chapter, you will know:

- The BODMAS scheme accommodates powers under the heading OF.
- Accordingly, powers have a high priority in the BODMAS scheme, and only BRACKETS are higher.
- To rearrange a power, we perform the inverse operation, which is generally a ROOT.

The mathematics of powers is discussed in Chapters 7–9. In this section, we merely continue our exploration of the trends in the previous section, and look at rearrangements involving power functions.

The first power functions are simple squares. Just looking at simple numbers ought to convince us that SQUARE and SQUARE ROOT are inverse functions. For example, $3^2 = 9$ and $\sqrt{9} = 3$. A little more thought reveals this rule to have a wider application, so $4^3 = 64$ and $\sqrt[3]{64} = 4$. In fact, we can deduce a more general rule still, which says that 'raising anything to its nth power' and 'taking the nth root' are inverse functions.

In general, we write:

$$a^n = b \text{ so } a = \sqrt[n]{b} \tag{5.1}$$

When taking a multiple root, such as $\sqrt[3]{x}$, the small figure nestling before the root sign tells us the level of the **root**. We should assume a *square* root if we see only a root sign with no small figure.

■ **Worked Example 5.7**

The dithionite ion (**II**) is an extremely powerful reducing agent, and readily reduces organic aromatic species such as tetracene (**I**) from the previous worked example.

$$^- \underset{O}{\overset{O}{\underset{\|}{\overset{\|}{S}}}} - \underset{O}{\overset{O}{\underset{\|}{\overset{\|}{S}}}} ^-$$

II

The rate law when reducing with dithionite is:

$$\text{rate} = k[\text{aromatic}]\sqrt{[(\textbf{II})]}$$

where k is a rate constant.

Looking at the right-hand side of the equation shows how the concentration $[(\textbf{II})]$ was first SQUARE ROOTED to form $\sqrt{[(\textbf{II})]}$, which was then multiplied by k [aromatic].

The first step when obtaining $[(\textbf{II})]$ is therefore to DIVIDE both sides of the equation by 'k [aromatic]'. We will treat k [aromatic] as a *compound variable*. We therefore perform the inverse operation, and DIVIDE both sides in a single step:

$$\frac{\text{rate}}{k\,[\text{aromatic}]} = \sqrt{[(\textbf{II})]}$$

Since the right-hand side is a SQUARE ROOT, to obtain $[(\textbf{II})]$ from $\sqrt{[(\textbf{II})]}$ we perform the inverse operation and SQUARE both sides:

$$\left(\frac{\text{rate}}{k\,[\text{aromatic}]}\right)^2 = \left(\sqrt{[(\textbf{II})]}\right)^2$$

The SQUARE of a SQUARE ROOT yields $[(\textbf{II})]$, so:

$$[(\textbf{II})] = \left(\frac{\text{rate}}{k\,[\text{aromatic}]}\right)^2 \quad\blacksquare$$

We can rearrange more complicated equations incorporating ROOT operators, provided we stick closely to the BODMAS rules.

■ **Worked Example 5.8**

The ionic conductivity Λ of ions moving through a solution relates to its ionic concentration c according to the **Onsager** equation:

$$\Lambda = \Lambda^0 - k\sqrt{[C]}$$

where k is a constant, and Λ^0 is the ionic conductivity when the concentration of ions is zero. Rearrange this equation to make concentration $[C]$ the subject.

The concentration term has been amended in three separate consecutive ways:

- This concentration $[C]$ was rooted (i.e. a function OF $[C]$) to yield $\sqrt{[C]}$.
- The root $\sqrt{[C]}$ was then MULTIPLIED by $-k$, to yield $-k\sqrt{[C]}$.
- Finally, Λ^0 was ADDED to the compound variable $-k\sqrt{[C]}$, to yield $\Lambda^0 - k\sqrt{[C]}$.

It often comes as something of a surprise to suggest that $\sqrt{[C]}$ was multiplied by *minus k*; it would have been equally valid to suggest that $\sqrt{[C]}$ was multiplied by $+k$ and only then was $k\sqrt{[C]}$ subtracted from Λ^0. Our approach here is intended to be easy and save time.

To make $[C]$ the subject of the equation, we first SUBTRACT Λ^0 from both sides:

$$\Lambda - \Lambda^0 = -k\sqrt{[C]}$$

Note how the minus sign persists on the right-hand side. Next, we note how $\sqrt{[C]}$ has been MULTIPLIED by a factor of $-k$, so we DIVIDE both sides by $-k$:

$$\frac{\Lambda - \Lambda^0}{-k} = \sqrt{[C]}$$

Finally we perform the inverse operation to SQUARE ROOT, and square both sides:

$$[C] = \left(\frac{\Lambda - \Lambda^0}{-k}\right)^2 \quad \blacksquare$$

Self-test 2

In each of the following, rearrange to make x the subject:

2.1 $y = x^2$

2.2 $y = -4x^2$

2.3 $y = x^2 + 7$

2.4 $y = 4(c - x^2)$

2.5 $y = c(x^2 + 1)$

2.6 $y = (x - a)^2$

2.7 $y = \sqrt{x} - 9$

2.8 $y = 5 \times \sqrt{x - v}$

2.9 $y = \sqrt{a - x}$

2.10 $y = \left(\dfrac{x + 1}{5}\right)^2$

Additional problems

The answers to some of these questions will be obvious. Nevertheless, determine the answer using the methodology outlined in this chapter.

5.1 By cancelling, show that the following two equations are the same:

(i) rate $= \dfrac{4\,k_2[\text{alcohol}]}{6}$ (ii) rate $= \dfrac{2\,k_2[\text{alcohol}]}{3}$

5.2 The molar mass of calcium carbonate is $100\,\text{g mol}^{-1}$. If the respective atomic masses of carbon and oxygen are 12 and $16\,\text{g mol}^{-1}$, determine the mass of 1 mole of calcium.

5.3 By cancelling, simplify the following **rate expression**:

$$\text{rate} = \frac{k_1\,k_2^2}{k_1\,k_2\,k_3}$$

5.4 All materials in solution move by diffusion. **Fick's second law** relates the distance diffused l over a period of time t, with the 'velocity of diffusion', known as the diffusion coefficient D:

$$l = \sqrt{Dt}$$

Rearrange the expression to make Dt the subject.

5.5 The **Einstein equation**:

$$E = mc^2$$

relates the energy released E when a mass m is converted entirely to energy. Rearrange the equation to make c the subject.

5.6 Consider the reaction ethane $\rightarrow$ ethane $+ H_2$ (a process known in the petrochemical industry as **cracking**).

The equilibrium constant for the cracking process is:

$$K = \frac{[H_2][ethene]}{[ethane]}$$

Rearrange the expression for the equilibrium constant, to make $[H_2]$ the subject.

5.7 The acceleration a of a molecular fragment inside a mass spectrometer is determined via the expression:

$$a = \frac{v - u}{t}$$

where the velocity changes from an initial value u to a different final value v during a time t. Rearrange the equation to make v the subject.

5.8 Assume that an atom of neon is spherical with a volume V of $4.85 \times 10^{-24}\,m^3$. If the equation relating radius r and volume V of a sphere is:

$$V = \frac{4}{3}\pi r^3$$

determine the atomic radius of a neon atom.

5.9 A chemist weighs out a mass of $10.2\,g$. The mass of the weighing boat is $0.250\,g$. If the molar mass of the compound is $34\,g\,mol^{-1}$, what amount of compound has been weighed out?

5.10 The equation relating a temperature in Fahrenheit F and a temperature in Celsius C is:

$$F = \frac{9(C + 32)}{5}$$

Rearrange this expression to make C the subject.

6

Algebra V

Simplifying equations: brackets and factorizing

Multiplying with brackets

By the end of this section, you will know:

- The correct way to multiply with a bracket is to MULTIPLY each term within the bracket by the factor positioned outside it.
- When two brackets are positioned side by side, we MULTIPLY them term by term.

We know from the BODMAS rule in Chapter 2 that brackets have the highest priority in a calculation. We now explore the best way to MULTIPLY them.

The simplest process is MULTIPLYING a single bracket with a factor or variable:

■ Worked Example 6.1

The amount of energy E needed to warm an amount of water n from an initial temperature $T_{initial}$ to a final temperature T_{final} is given by the expression:

$$E = nC[T_{final} - T_{initial}]$$

where C is the heat capacity of the water. Multiply out the bracket.

We will treat n and C as a compound variable, nC. We MULTIPLY nC by each temperature in turn:

$$E = nC[T_{final} - T_{initial}]$$

The first multiplication step (**1**) yields $nC \times T_{final}$. The second multiplication step (**2**) yields $nC \times T_{initial}$. Combining the two terms:

$$E = nC\,T_{final} - nC\,T_{initial}$$

$$\quad\quad 1 \quad\quad\quad 2$$

Sometimes an equation incorporates several brackets, positioned together. As usual, the absence of an operator between the brackets means that we MULTIPLY them together. ■

We often say we **multiply out** the brackets.

■ Worked Example 6.2

The van der Waals equation is one of the better ways to take account of the forces which cause the ideal-gas equation to fail:

$$\left(p + \frac{n^2a}{V^2}\right)(V - nb) = nRT$$

MULTIPLY together the two brackets on the left-hand side.

In a similar way to MULTIPLYING a single bracket with a factor, we here MULTIPLY term by term. In effect, we treat the first bracket as though it comprised two factors, p and $n^2a \div V^2$. We first MULTIPLY both of the terms in the second bracket by p, then MULTIPLY both terms in the second bracket by $n^2a \div V^2$. Finally, we add together the two results.

First we MULTIPLY the left-hand bracket with p:

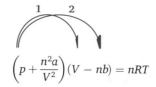

$$\left(p + \frac{n^2a}{V^2}\right)(V - nb) = nRT$$

The results of the two MULTIPLICATION steps are:

Step 1: pV and **Step 2**: $-pnb$

Then we MULTIPLY the right-hand bracket with $n^2a \div V^2$:

$$\left(p + \frac{n^2a}{V^2}\right)(V - nb) = nRT$$

3 4

The results of the two MULTIPLICATION steps are:

Step 3: $p \times \dfrac{n^2a}{V^2}$ **Step 4**: $-nb \times \dfrac{n^2a}{V^2}$

We then sum the terms : $(pV) - (pnb) + \left(p \times \dfrac{n^2a}{V^2}\right) - \left(nb \times \dfrac{n^2a}{V^2}\right)$

1 2 3 4

We will learn later how to tidy up this answer. ∎

> Each bracket contains two terms. The final result will therefore comprise four terms, because $2 \times 2 = 4$.

> In general, multiplying out the brackets $(a+b)(c+d)$ yields $ac + ad + bc + bd$.

Self-test 1

MULTIPLY out the following expressions:

1.1 $p(a+b)$

1.2 $bp(x-y)$

1.3 $(a+1)(c+2)$

1.4 $(a+b^2)(a^2-b)$

1.5 $g(H+2)(J+3)$

1.6 $(J+2)(J-2)$

1.7 $(\frac{1}{2} + 4v)(3 - a/c)$

1.8 $(1/c - 1/d)(e+5)$

Squaring a bracket: the 'perfect square'

By the end of this section, you will know:

- When we square a bracket, we multiply it with itself to form a 'perfect square'.
- Squaring the bracket $(a+b)$ generates $a^2 + 2ab + b^2$.

We employ the same methodology when we SQUARE a bracket such as $(a+b)^2$ as multiplying together two different brackets:

$$(a+b)^2 = (a+b) \times (a+b)$$

so:

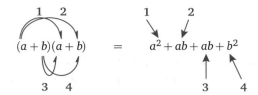

therefore:

$$(a + b)^2 = a^2 + 2ab + b^2 \qquad (6.1)$$

We generally consider that squaring a bracket is a special case, because we can simply write the answer via eqn (6.1) and thereby obviate the necessity to multiply out each bracket separately.

■ **Worked Example 6.3**

Show that $(x + 4)^2$ follows the general scheme in eqn.

Firstly we consider the x factor:

$$(x + 4)(x + 4) \qquad = x^2 + 4x$$

Secondly, we consider the factor of 4:

$$(x + 4)(x + 4) \qquad = 4x + 16$$

so:

$$(x + 4) \quad = x^2 + 4x + 4x + 16$$

so:

$$(x + 4)^2 = x^2 + 2 \times 4x + 4^2 = x^2 + 8x + 16$$

Self-test 2

Multiply out each of the following squares, and show how the result is the same as that generated with eqn (6.1):

2.1 $(a + b)^2$ 2.5 $(2x - y)^2$

2.2 $(x - y)^2$ 2.6 $(ax - by)^2$

2.3 $(-x + c)^2$ 2.7 $(5 - y^2)^2$

2.4 $(-x - y)^2$ 2.8 $(1 - 4y)^2$

Factorizing simple expressions

By the end of this section, you will know:

- Factorizing is one of the simplest ways to simplify an equation.
- Factorizing is often the inverse process of multiplying brackets.

When we **factorize**, we look at an equation and look for factors in common. For example, if we have $ab + ac$, then both terms contain the common factor a. Similarly, if we have $4x + 8$, then 4 is a factor in each of the two terms, because 4 goes into both 4 and 8. The factors can be more complicated, so the expression $4x^2 + 2x$ has $2x$ as a factor.

> A **factor** is a number or algebraic symbol with which we multiply something else.

■ Worked Example 6.4

The energy E of a gas particle has two components, both kinetic and potential energies. The potential energy is mgh, and the kinetic energy is $\frac{1}{2}mv^2$, where m is mass, g is acceleration due to gravity, h is height, and v is velocity. Factorize the expression:

$$E = mgh + \tfrac{1}{2}mv^2$$

Both terms contain m as a factor. In effect, then, we could write the expression as:

$$E = (m \times gh) + \left(m \times \tfrac{1}{2}v^2\right)$$

This expression is identical in every way to that above except our choice of notation style: we have separated the factor of m from each of the two terms. Since m is common to both, we then write:

$$E = m\left(gh + \tfrac{1}{2}v^2\right)$$

This is the factorized expression. In other words, we have collected together the two common m factors. If we were to multiply out this bracket, we would regain the original expression. ■

■ Worked Example 6.5

Dalton's law says the total pressure exerted by a mixture of gasses p_{total} equals the sum of the partial pressures of the gases. If a mixture comprises argon, nitrogen, and hydrogen at pressure of $p(Ar) = ap^{\ominus}$, $p(N_2) = bp^{\ominus}$, and $p(H_2) = cp^{\ominus}$, use Dalton's law to add up these pressures, then factorize them. ($p^{\ominus}$ here is standard pressure.)

Dalton's law says:

$$p_{total} = p(Ar) + p(N_2) + p(H_2)$$

Substituting for the numbers:

$$p_{total} = ap^{\ominus} + bp^{\ominus} + cp^{\ominus}$$

Factorizing, we say:

$$p_{total} = (a + b + c) \times p^{\ominus}$$

It would have been equally correct to have written $p_{total} = p^{\ominus}(a + b + c)$.

Self-test 3

Factorize each of the following:

3.1	$a + ab$	**3.6**	$a^2 + 4a$
3.2	$2x - 2y$	**3.7**	$a^2 + 3a + 6ab$
3.3	$ab + 2ac + ad$	**3.8**	$xy - x^2 + 2x$
3.4	$4a + 2ac + 6ab$	**3.9**	$x^3 + 6x^2 + 4x$
3.5	$a^2 + ab$	**3.10**	$(x + 4)^2 + (2x - 3)^2 - (25 + x)$

Factorizing the difference between two squares

By the end of this section, you will know:

- We call equations of the sort $x^2 - y^2$ 'the difference between two squares'.
- The difference between two squares $x^2 - y^2$ factorizes as $(x - y)(x + y)$.

Possibly the simplest factorization starts with equations of the type $a^2 - b^2$, which factorize to $(a - b)(a + b)$. We will call it the **difference between two squares**.

■ **Worked Example 6.6**

Consider the right-angled triangle below:

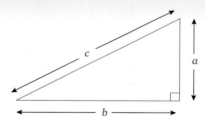

Pythagoras's theorem says that $b^2 = c^2 - a^2$. Factorize this equation.

The right-hand side of Pythagoras's equation represents the difference between two squares, so as:

$$b^2 = c^2 - a^2 = \text{the difference between two squares}$$

then:

$$b^2 = (c - a)(c + a) \quad ■$$

Sometimes the second term when dealing with the difference between two squares does not immediately look like a square. In this case, we rewrite the difference between the two squares somewhat, and say:

$$(x^2 - a) = (x - \sqrt{a})(x + \sqrt{a})$$ (6.2)

■ **Worked Example 6.7**

Factorize the expression $y = x^2 - 9$.

As the right-hand side represents the difference between two squares, the factors will be $(x - \sqrt{a})$ and $(x + \sqrt{a})$, where each $\sqrt{a}$ term here is the square root of 9. The equation factorizes as:

$$y = (x - 3)(x + 3) \quad ■$$

Self-test 4

Factorize the following differences between two squares, using eqn (6.2):

4.1 $x^2 - b^2$	4.5 $x^2 - 4$
4.2 $c^2 - p^2$	4.6 $a^2 - 81$
4.3 $\alpha^2 - \beta^2$	4.7 $d^2 - 10{,}000$
4.4 $\omega^2 - \pi^2$	4.8 $e^2 - 5.32$

Factorizing quadratic equations

By the end of this section, you will know:

- A quadratic equation has four terms, hence its name, and has the form $0 = ax^2 + bx + c$, where a, b, and c are factors (or 'coefficients').
- If $a = 1$, and b and c are whole numbers, we factorize the equation $0 = x^2 + bx + c$, saying $(x + A)(x + B)$, where $b = (A + B)$ and $c = A \times B$.
- When $a \neq 1$, we may have to factorize by a process of trial and error.

The equations we factorized above were relatively simple. For example, a perfect square only has a single term containing x. We look now at more complicated equations, in which there is more than one term in x, and which cannot straightforwardly be factorized.

■ **Worked Example 6.8**

What are the factors of the equation $y = x^2 + 3x + 2$?

Altogether, this equation contains four terms, so we call it a **quadratic equation**, following the Latin convention which says that *quad* always means four.

Before factorizing the expression, quite reasonably we assume it must have come from somewhere—it was formed when two brackets were MULTIPLIED together. Our task is to work backwards, asking the question, 'What was in those two brackets?'

Next, we consider the general case where we multiply together the two brackets and compare it with the equation in the question:

general case $y = (x + A) \times (x + B) = x^2 + \quad (A + B)x \quad + AB$

special case $y = \qquad\qquad\qquad\qquad x^2 + \quad 3x \qquad + 2$

The Latin prefix **quad** always means 'four', so a *quadrangle* is an area bounded by four walls, and a *quad bike* has four wheels.

By comparing the right-hand side of the equation in the question (x^2+3x+2) with the general equation $(x^2+(A+B)x+AB)$, we see how $(A+B)=3$ and $AB=2$. Therefore:

$$A = 1 \text{ and } B = 2$$

so:

$$y = x^2 + 3x + 2 \text{ factorizes to } y = (x+1)(x+2) \; \blacksquare$$

We could equally have said $A=2$ and $B=1$, but the answer would have been the same, because multiplying brackets is **associative** (see Chapter 3).

Occasionally, this exercise can be quite difficult:

■ **Worked Example 6.9**

What are the factors of the equation $y=x^2+22x+40$?

Using the relationships immediately above, the coefficient 40 is the product of A and B, while the coefficient of 22 is the sum of A and B. In an example such as this, the identity of A and B may not be immediately apparent. We therefore need a method to determine their values.

Strategy:

(i)   We 'brainstorm', looking for pairs of numbers whose product is 20. Obvious examples involving only whole numbers are 1×40, 2×20, 4×10, and 5×8.

(ii)   We look more closely at the pairs of numbers generated in step (i). Specifically, we ask what is their sum. Remember: we are looking for a pair with a sum of 22.

Product	Sum
1×40	41
2×20	22
4×10	14
5×8	13

At this stage, it should be clear that only 2 and 20 fulfil both criteria. Accordingly, these are the roots of this quadratic. We say $x=2$ and $x=20$, so $(x-2)=0$ and $(x-20)=0$. Therefore:

$$0 = (x-2)(x-20) \; \blacksquare$$

Self-test 5

Factorize the following quadratic equations by *inspection*: use the method above.

5.1 x^2+4x+3	5.6 $x^2+25x+100$
5.2 x^2+5x+6	5.7 x^2+6x+9
5.3 x^2+5x+4	5.8 $x^2+15x+56$
5.4 x^2+7x+6	5.9 $x^2+7x+10$
5.5 $x^2+9x+20$	5.10 $x^2+8x+15$

Factorizing with 'the formula'

By the end of this section, you will know:

- 'The formula' will solve any quadratic equation (except those without a real answer).
- We factorize quadratic equations with 'the formula' when the coefficients a, b, and c are not integers.
- To use 'the formula', we sometimes need first to rearrange in order to start with an equation of the form '$0 = \ldots$'.

If we cannot factorize a quadratic equation using simple patterns (as above), we employ 'the formula'. Provided the quadratic has the form $0 = ax^2 + bx + c$, we say:

$$x = \frac{-b \pm \sqrt{b^2 - 4ac}}{2a} \qquad\qquad (6.3)$$

where the symbol $\pm$ means **plus or minus**. In effect, the $\pm$ sign tells us there are *two* possible answers that will correctly satisfy the equation. We might have expected this result because a quadratic equation should factorize to form *two* brackets.

■ **Worked Example 6.10**

Use the formula to factorize $0 = 3x^2 + 14x + 8$.

Strategy:

(i)   We identify the coefficients, a, b, and c.

(ii) We use the formula in eqn (6.3) to determine the two values of x that satisfy the quadratic equation.

(iii) Knowing the values of x, we derive two brackets as the factors of the equation.

(i)   We start by identifying the factors: the number MULTIPLYING x^2 is $a = 3$, the number MULTIPLYING x is $b = 14$, and the free-standing number at the end is $c = 8$.

(ii) We then insert these coefficients into eq. (6.3):

$$x = \frac{-14 \pm \sqrt{14^2 - 4 \times 3 \times 8}}{2 \times 3}$$

which becomes:

$$x = \frac{-14 \pm \sqrt{196 - 96}}{6}$$

The value of the square root is $\sqrt{100}$, which is clearly 10. As a result of the $\pm$ sign, factorizing with the formula yields *two* answers:

$$x = \frac{-14 \pm 10}{6}$$

$$x = \frac{-14 + 10}{6} \qquad \text{or} \qquad x = \frac{-14 - 10}{6}$$

$$x = -\frac{4}{6} = -\frac{2}{3} \qquad \text{or} \qquad x = -\frac{24}{6} = -4$$

We call the two values of x which **satisfy** this quadratic its roots. The **roots** of the equation $0 = 3x^2 + 14x + 8$ are $-2/3$ and -4.

(iii) To obtain the factors of the equation, we rearrange the formulae of these **roots**:

if $x = -4$ then adding 4 to both sides yields $0 = (x + 4)$

and $x = -\frac{2}{3}$ then adding $\frac{2}{3}$ to both sideis yields $0 = \left(x + \frac{2}{3}\right)$

We need to modify the second bracket, so multiply throughout by 3, giving: $(3x + 2) = 0$.

Therefore, the equation $0 = 3x^2 + 14x + 8$ factorizes to: $(x + 4)(3x + 2)$. ∎

■ **Worked Example 6.11**

The leaves of a plant never grow randomly around the stem, but grow in a regular spiral with successive leaves forming at fixed angles (see Fig. 6.1). This fraction of a circle between successive leaves represents the proportion of a circle that fulfils the equation $x^2 = x + 1$. What is x?

The angle 137° is common throughout nature. This proportion of a circle is called the **Golden Ratio** and given the Greek letter tau τ. The ratio of the circle sectors is $1 : \tau$, where τ fulfils the equation $\tau^2 = \tau + 1$.

We first rearrange slightly, so instead of saying $x^2 = x + 1$, we say $0 = x^2 - x - 1$. The coefficients are clearly $a = 1$, $b = -1$, and $c = -1$.

Figure 6.1 The angle between successive leaves of *Fibonacci phyllotaxis* is close to the Golden Ratio of about 137.5°. (Figure reproduced by kind permission of Professor Pau Atela, Mathematics Department, Smith College, USA).

Next, we insert coefficients into the formula (eqn (6.3)):

$$\frac{1 \pm \sqrt{1^2 - (4 \times 1 \times -1)}}{2 \times 1}$$

so:

$$\frac{1 \pm \sqrt{1+4}}{2}$$

and:

$$\frac{1 \pm \sqrt{5}}{2}, \text{ where } \sqrt{5} = 2.236$$

There will be two roots to the equation:

$$x = \frac{1 \pm 2.236}{2}$$

$$x = \frac{1 + 2.236}{2} \qquad \text{or} \qquad x = \frac{1 - 2.236}{2}$$

$$x = \frac{3.236}{2} = 1.62 \qquad \text{or} \qquad x = -\frac{1.236}{2} = -0.618$$

so the equation $0 = x^2 - x - 1$ has the roots $x = 1.62$ and $x = -0.618$.
The equation factorizes to $(x - 1.62)(x + 0.618)$. ∎

The Golden Ratio corresponds to the first root, $x = 1.62$. (Being negative, the second root does not correspond to a real, physical situation: a negative angle is not useful.)

■ **Worked Example 6.12**

Triflouroethanoic acid (**I**) ionizes in water to form one anion and one solvated proton:

The ionization reaction does not proceed to completion; rather, a proportion x ionizes while the remainder persists as a neutral, covalent molecule, according to the equilibrium-following constant:

$$K = \frac{[H^+][\text{anion}^-]}{[\text{un-ionized acid}(I)]}$$

At equilibrium, the concentrations of $[H^+]$ and $[\text{anion}^-]$ are both x, and the concentration of un-ionized (**I**) will be $(1 - x)$.
   What is the value of x if $K = 0.3$?

Strategy:

(i)   We insert numbers into the equation for K.

(ii) We rearrange the equation to generate a quadratic equation.

(iii) We solve the quadratic equation using the formula.

Insert these lengths into Pythagoras's equation in Worked example 6.6, factorize the data, and hence calculate the length c.

6.10 The dissociation constant K of ethanoic acid (**II**) is 1.75×10^{-5}. Using the same methodology as that in Worked example 6.12, calculate the proportion x of the acid that is dissociated if the concentration of the acid is 0.1 mol dm^{-3}.

II

Graphs I

Introducing pictorial representations of functions

Introducing the concepts underlying graphical representations

By the end of this section, you will know:

- A graph is a pictorial representation of data.
- The names of the x and y axes.
- The x axis is the horizontal axis, and represents the **controlled** variable.
- The y axis is the vertical axis, and represents the **observed** variable.

As the old proverb says, 'A picture is worth a thousand words.'

Scientists often want to know whether a relationship exists between a pair of variables. It's a more advanced version of the child who says, 'What happens if I press this button?' Such scientists will probably call their investigations 'an experiment'.

The scientist's investigation is more advanced than the child's for several reasons. Firstly, because the scientist is more responsible, so he or she does not try sticking a finger into an electrical socket, but actually thinks about the experiment before starting it. Secondly, the scientist knows that variables come in two general types, both 'controlled' and 'observed', as follows. Imagine the scientist designs an experiment to see if a radio gets louder while adjusting the volume control. The two variables are 'loudness' and 'the position of the volume knob'. The scientist deliberately tweaks the volume control to see whether the radio output changes. We give the name **controlled variable** to the adjustments to the volume knob, because we dictate its position. We give the name **observed variable** to the loudness resulting from adjusting the controlled variable (the volume knob).

It's easy to turn the knob and measure how loud the radio is; it is more difficult to decide beforehand what loudness we want and then find the necessary position of the volume knob: we would probably need to turn the knob back and forth until we obtain the loudness we want. This 'fine tuning' illustrates how a scientist should design an experiment to enable the magnitude of the controlled variable to span a wide range: it is generally more difficult if we want swap the variables, in this case deciding what the loudness should be and seeing where to position the volume knob.

So the scientist performs a series of experiments, varying the controlled variable (say) a dozen times, and obtaining a dozen values of the observed variable. These data help the scientist to decide if a relationship does relate the two variables. The scientist will probably first write the data in a table and ask, 'When I turned the volume knob, do the data show that the loudness changed?' A mere glance at the numbers will show if changing one variable causes the other to change. If so, we say a **relationship** exists. Some scientists might say a **correlation** exists, which means the same thing.

Few scientists remain content with knowing a relationship exists: they usually want to know the *nature* of that relationship. While a table of data demonstrates the existence of a relationship, it is generally a lousy way of ascertaining the nature of the relationship.

Before we look at the ways of determining the mathematical nature of a relationship, we need to introduce two more terms. A **qualitative** measurement tells us the qualities of the things investigated. Looking at a table of data tells us the result that loudness and the position of a radio's volume knob are related. Other simple

Remember:

The word **data** implies a plural; the singular is **datum**.

qualitative statements might be, 'This colour is brighter than that one', 'This reaction gets hotter than that one', or 'This student is the tallest in the class.' Each refers to a genuine difference in the quality of a thing, but none suggests a numerical value. If we want to put a number to the differences, perhaps saying, 'I travelled twice as fast as you', we say more than a qualitative statement: we say something **quantitative**. Quoting the value of an enthalpy, entropy, or an *emf* is a quantitative statement.

Self-test 1

Decide which of the following statements are qualitative and which are quantitative:

1.1 $\Delta H_{\text{(reaction)}} = 12\,\text{kJ mol}^{-1}$

1.2 Your flask is bigger than my beaker

1.3 The temperature rose faster during the second experiment

1.4 $\lambda = 450\,\text{nm}$

1.5 We are late for the class

1.6 The temperature was raised by 2.3 °C

Quantitative measurements are generally more difficult to perform than qualitative measurements, so we generally start a chemical investigation by assessing qualitative relationships and only then do we make quantitative statements—it would be silly to quantify if no relationship exists. For example, we first perform a reaction and ascertain whether a reaction has occurred, and what is formed. Only then is it sensible to calculate the chemical yield. The yield of a reaction cannot be calculated if we don't even know what has been formed.

Drawing a **graph** is one of the best ways to discern the quantitative relationship between the observed and controlled variables. The most common graphical method is that attributed to the French philosopher and scientist René Descartes (1596–1650). His name explains why we say many parameters relating to graphs are **Cartesian**.

We always start a graph by drawing the controlled variable along the horizontal axis. We generally call this axis the *x* axis. Its more scientific name is the **abscissa**. We draw the data concerning the observed variable along the vertical axis, which we call either the *y* axis or the **ordinate**. We call the intersection of the two axes the origin, because the two axes originate from this point.

> ....
> The *x* **axis** is the horizontal axis while the *y* **axis** is the vertical axis. One way to remember which axis is which is to say, 'an eXpanse of road goes horizontally along the *x* axis', and 'a Yo-Yo goes up and down the *y* axis'.
> ....

> ....
> The word **abscissa** comes from the Latin *absindere abscissa*, which means in this context 'to cut'. The word **ordinate** again comes from the Latin, and represents an abbreviation of *linea ordinato applicata* – 'a line drawn parallel'.
> ....

Plotting the graph

By the end of this section, you will know:

- The purpose of drawing a graph is to discern whether two variables are related—we look for a *correlation*.
- How to plot a graph.
- How to label the graph's two axes.

To **plot** a graph, we first obtain complementary pairs of data: that is, each value of the controlled variable *x* is associated with a value of the observed variable *y*. We may call them (x_1, y_1), (x_2, y_2), etc., where the subscripted numbers indicate perhaps

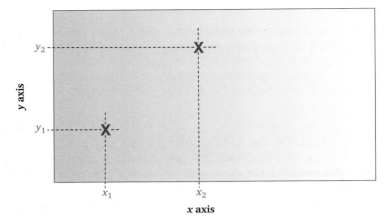

Figure 7.1 We draw a graph by drawing a vertical line through each value of x and a horizontal line through each value of y. We draw a small cross x to represent the point where the two lines intersect, and then draw the line of best fit through the crosses.

the order in which we obtain the data. Notice that we always cite the controlled variable first.

Next, for each pair, we draw a vertical line through the value of x and a horizontal line through the value of y. We indicate the point where these 'guidelines' intersect with a small cross x, as represented in Fig. 7.1. It does not matter how large or small we draw the crosses, provided they are legible, consistent, and neat. We lightly draw these guidelines in pencil when we first attempt such graphs, but we no longer need to draw them physically after a little practice. We see them 'in our mind's eye' as a prelude to drawing a physical cross. Some people draw the guidelines, draw a cross, and then rub out the guidelines to make the graph neater.

Having plotted the data, we then label the resulting graph. Half way along the x axis scale we write a description (involving both words and symbols) to indicate the nature of the controlled variable. We position the label beneath the rows of numbers below the line. We position the label for the y axis half way up the axis and slightly to the left of the numbers making up the y axis scale.

■ **Worked Example 7.1**

Hydrogen peroxide H_2O_2 decomposes in the presence of excess cerous ion Ce^{III} to form ceric ion Ce^{IV}:

$$2H_2O_2 + 2Ce^{III} \rightarrow 2H_2O + Ce^{IV} + O_2$$

The following data were obtained at 298 K, and show how the concentration of H_2O_2 decreases with time t. Plot the data with time as the controlled variable and the concentration $[H_2O_2]_t$ as the observed variable:

Time, t/s	2	4	6	8	10	12	14	16	18	20
$[H_2O_2]_t/$ mol dm^{-3}	6.23	4.84	3.76	3.20	2.60	2.16	1.85	1.49	1.27	1.01

....

The subscripted 't' printed on the outer right-hand side of the concentration bracket tells us these concentration change with time t after the reaction starts. In this same nomenclature, the concentration at the start of the reaction is $[H_2O_2]_0$.

....

Methodology:

(i)   We first decide which are the controlled and which the observed variables.

(ii) Having decided, we plot the controlled variable on the vertical y axis and the controlled variable on the horizontal x axis.

There are two ways to decide which is the controlled and which the observed variables.

(i)   When compiling a table of data, it is common practice to write the controlled axis on the top row and the observed data as lower rows. This convention instructs us that time is the controlled variable. Concentration is therefore the observed variable. This method is not always safe—for example, the person compiling the table might not have known the convention.

(ii) So the best way to decide which is the controlled and which the observed variable is to imagine ourselves performing the experiment ourselves. A moment's thought suggests it would be very difficult indeed to dictate the concentration of the peroxide, because it keeps changing. We can't wait until the concentration reaches a certain value, and then notice the time. It would be too fiddly. On the other hand, it's quite easy to imagine ourselves with a stop-watch in one hand, deciding when to take each reading. ■

....
As corroboration, we notice how the time increments are evenly spaced and the concentrations are not. Normally, only a controlled variable will be evenly spaced in this way.
....

Knowing that the time is the controlled variable, we plot times t along the horizontal x axis. 'Concentration' represents the observed variable, so we plot $[H_2O_2]_t$ up the vertical y axis; see Fig. 7.2.

Self-test 2

Plot the following graphs. In each case, use values of x from -4 to $+4$.

2.1 $y = 5x + 3$ 2.5 $y = 0.3x + 0.2$

2.2 $y = 12x - 2$ 2.6 $y = x$

2.3 $y = -4x + 1$ 2.7 $y = 0$

2.4 $y = 0.2x$ 2.8 $x = 2$

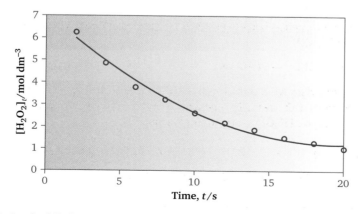

Figure 7.2 Graph of (i) the concentration $[H_2O_2]_t$ as the observed variable, plotted on the y axis, and (ii) time as the controlled variable, plotted on the x axis.

The correct appearance of a graph

By the end of this section, you will know:

- A graph must always be labelled, and where to place the label.
- A graph should fill the page.
- A straight-line graph ought to be inclined at an angle close to 45°.

Position on a page The data should fill the page as far as possible. A graph that occupies one small part of the page is not only wasting paper, but also contributes towards error. If the graph is squeezed into a small space, we are likely to experience errors in reading from it.

Labelling Both axes on a graph must be labelled. The correct position is indicated on the graphs in Figs 7.1 and 7.2, i.e. just below the x axis and just to the left of the y axis.

Look at each label in Fig. 7.2. The vertical scale says more than 'concentration of analyte', and the horizontal scale says more than 'time t'. Neither description is complete until we indicate the *units*. The label is incomplete without the respective units. The correct way to include the correct units is very straightforward. We write this as 'quantity ÷ units', so a concentration axis might be labelled 'concentration, c/mol dm^{-3}' and a temperature axis as 'temperature, T/K'.

There exists a simple reason why we divide by the respective units: we can only plot *numbers*. If this were not the case, one side of a temperature axis would be hot while the other would be cold. Unfortunately, we did not measure mere numbers before plotting the graph: we measured a temperature, etc. As an example, let the first temperature we measure be 320 K. We will plot the number as '320'. To do so, we take the algebraic phrase, 'temperature $T = 320$ K'. If we divide both sides of this little equation by the unit of K, we obtain 'temperature T/K $= 320$'. We write the left-hand side of the equation as the axis label, and plot the right-hand side as the data points on the graph.

The angle of the line If a graph is linear (see below), the line should be positioned at an angle close to 45° to either axis. If the line is positioned such that it is nearly horizontal or nearly vertical on the page, we will experience difficulty in reading from it. For example, look at the graph in Fig. 7.3. The spacing between the values of the controlled variable x_1 and x_2 is wide, and therefore easily read from the graph. Conversely, because the line was drawn at such a shallow angle, the respective

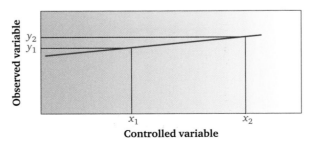

Figure 7.3 It is not possible to read from the y axis with any accuracy if the graph is drawn nearly horizontally.

values of the observed variables y_1 and y_2 are so close together they cannot be easily differentiated. Accordingly, if we were to use Fig. 7.3 as a calibration curve, it is likely that values we read from the graph would be in error.

In a similar way, if the graph were drawn with a straight line that is nearly vertical, the errors associated with readings x_1 and x_2 would again be high. For these reasons, we minimize errors when drawing graphs by drawing the line close to an angle of 45°.

If a graph is curved, we should manipulate the values of x and y in such a way that the portion of interest is neither too shallow nor too steep. And, above all else, we should aim to fill the page as far as practicable.

Types of graph

By the end of this section, you will know:

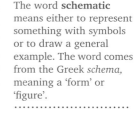

- Looking at a graph will often tell us the general type of correlation between the two variables.
- One of the common causes of graphs showing scatter is the un-compensated existence of *compound variables*.

Next, we look at several types of graph, each demonstrating a different kind of relationship between the observed and controlled variables. In common parlance, we talk about plotting the observed variable 'against' the controlled variable. It is bad practice to cite the controlled variable first; people will not expect us to cite the controlled variable first. If we plotted a graph to show the loudness of a radio as a function of the volume knob's position, we would say, 'A graph of loudness against knob position.' It would be wrong to say, 'A graph of knob position against loudness.'

Figure 7.4 shows a schematic of a straight-line graph, which passes through the origin. We would still call it a 'straight-line graph' even if the line had not passed through the origin. Figure 7.4 tells us that when we vary the controlled variable x, the observed variable y changes in direct proportion. A glass of orange cordial affords an obvious example of such a situation: the intensity of its colour increases in linear proportion to the concentration of the cordial concentrate, according to the Beer–Lambert law. The graph in Fig. 7.4 goes through the origin because no orange colour is discernible before we add cordial to the water (i.e. its concentration is zero before we add any coloured cordial). We draw a bold straight line through the data if they fall on a perfect straight line, or the best straight line through the data if they display any scatter.

....

The word **schematic** means either to represent something with symbols or to draw a general example. The word comes from the Greek *schema*, meaning a 'form' or 'figure'.

....

....

Linear means line, from the Latin *linea*, meaning 'a line'. Scientists generally take the word 'linear' to mean a *straight* line.

....

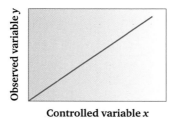

Figure 7.4 A graph of observed variable (along the y axis) against controlled variable along the x axis: a simple linear proportionality, so $y = $ constant $\times x$.

Figure 7.5 shows a different type of relationship. Firstly, although the graph is not straight it demonstrates that a relationship does indeed exist between the observed variable *y* and the controlled variable *x*, because the data display a clear *trend*. This graph tells us that the observed variable *y* increases at a faster rate than does the controlled variable *x*. A simple example might be the rate at which a reaction proceeds (as the observed variable) as we raise the reaction temperature (as the controlled variable). The bold line again represents the best line through the experimentally determined data. (We look at non-linear graphs in more detail in later chapters.)

The graph in Fig. 7.6 is another straight-line graph, but this time it is horizontal. In other words, whatever we do to the controlled variable *x*, the observed variable *y* remains unchanged. We say, 'the variable *y* is not a function of *x*': we discern this lack of a correlation because changing *x* does not cause *y* to change. A simple example of such a graph would be the yield of a reaction (as *y*) against the time of day (as the controlled variable, *x*). In the absence of other chemicals and variables, the yield will be same whether we react the chemicals in the morning, afternoon, or evening.

We must be careful before we say that no correlation exists. It is easy to lose sight of a genuine correlation if we plot a graph with too insensitive a *y* axis scale. For example, if the values of *y* vary between a minimum of 50.1 and a maximum of 50.2, then plotting a graph using silly *y*-scale limits such as 0 and 100 would yield a graph looking just like that in Fig. 7.6. In fact, the lack of a correlation is *apparent* rather than real: if we re-plotted the data using a *y*-scale range of 50.1 and 50.2, we would generate a graph looking much like Fig. 7.4 or Fig. 7.5.

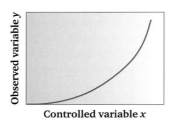

Figure 7.5 A graph depicting a situation in which the observed variable *y* is not a simple function of *x*, although relationship clearly does exist: a good example might be the rate of a chemical reaction (as *y*) against the temperature (as *x*).

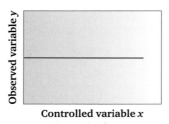

Figure 7.6 A graph depicting a situation when the variable *y* is independent of variable *x*: the yield of a chemical reaction (as *y*) against the time of day (as *x*).

■ **Worked Example 7.2**

The data below relate to the isomerization of 1-butene (**1**) to form *trans* 2-butene (**1I**):

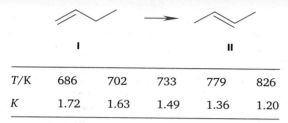

T/K	686	702	733	779	826
K	1.72	1.63	1.49	1.36	1.20

The equilibrium constants of reaction K clearly depend on the temperature. To demonstrate the importance of a choosing axis scales, we plot the data twice:

(i)   A y scale extending from 1.20 to 1.80, and an x axis extending between 650 and 850 K.

(ii)   A y scale extending from 0.00 to 10 and an x axis extending from 0 to 900 K.

Look at the graphs in Fig. 7.7. The upper graph (a) is vastly superior because:

• The data fill the page.

• The data occupy the whole of both axes, so it is easier to read data from the graph. ■

When plotting these data with Excel™, the second set of axes is the default choice of scales.

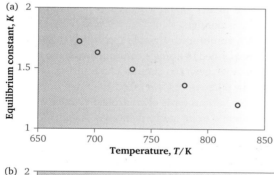

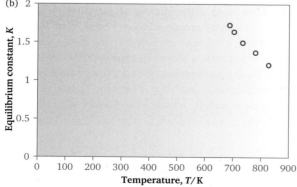

Figure 7.7 Graphs to demonstrate the need for a sensible scale on the y axis: (a) sensible axes, and (b) foolish axes.

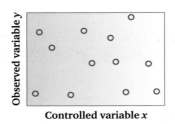

Figure 7.8 A graph depicting the situation in which no relationship is apparent between the observed variable *y* and the controlled variable *x*, although y does vary.

We need to note:

- As a good generalization, we plot the data of a straight-line graph to fill the page.
- The straight line should reside at an angle of about 45°.

The graph in Fig. 7.8 shows another common situation: the data indicate no straightforward relationship. In fact, Fig. 7.8 might be telling us that no relationship exists at all. The magnitude of the controlled variable *x* does not have any bearing on the observed variable *y*. We say the observed variable *y* is *independent* of the controlled variable *x*. There is no correlation. Nevertheless, we see a range of *y* values as *x* is varied.

Several possibilities could explain the apparent lack of a correlation. Firstly, perhaps the value of *y* is completely random, in which case repeating the experiment using the same values of controlled variable *x* would generate wholly different values of the observed variable *y*. We would then conclude that our data are irreproducible. Faulty equipment represents one of the simplest causes of irreproducible data, and poor control over the variables is another.

Secondly, perhaps *x* is a compound variable (see Chapter 3). We initially assumed the observed variable *y* responded to a *single* controlled variable *x*, but the data suggest our original analysis may have been too simplistic and in fact several variables dictate the observed value of *y*. In other words, we failed to design our experiment properly.

Perhaps we can call these various types of controlled variable x_1, x_2, etc. An everyday example might be a student's exam performance as *y* and IQ as x_1. The lack of a correlation suggests that while IQ is important, other variables also impinge on the magnitude of the student's exam result, such as commitment as x_2 and state of health— both mental and physical—as x_3. A more realistic choice of controlled variable *x* here therefore might be, 'How good is the student at chemistry on the day of the exam?' which is certainly a complicated function of x_1, x_2, and x_3. That we did not discern more than one contributory factor suggests we failed to design our experiment properly.

If many variables participated, a more rigorous approach would be to keep all but one of them constant during an experiment, and to ascertain how the *single* controlled variable x_1 affects the observed variable *y*, then how x_2 affects *y*, and so on. A good example is the way that pressure *p*, volume *V*, and temperature *T* interrelate according to the ideal-gas equation:

$$pV = nRT$$

where *R* is the gas constant and *n* is the amount of gas. If all three variables change at once, then a graph of temperature *T* as *y* against either pressure *p* or volume *V* as *x* will appear scattered. To ensure a linear graph, we must calculate the compound variable $p \times V$ and plot it as *x*.

Self-test 3

We inflate a car tyre. The pressure p inside the tyre increases concurrently with changes in the internal pressure p and the internal temperature T, according to the data below:

Pressure, p/Pa	10 000	20 000	30 000	40 000	50 000
Volume, V/m³	0.4	0.1	0.5	0.3	0.05
Temperature, T/K	481	240	1804	1443	301

The data refer to 1 mole of an ideal gas. Plot the following three graphs:

3.1   Temperature T (as y) against 'pressure p' (as x).

3.2   Temperature T (as y) against 'volume V' (as x).

3.3   Temperature T (as y) against 'pressure × volume', i.e. $p \times V$ (as x).

Note how the x axis differs, but the y axis of each is the same.

Incidentally, the exercise in Self-test 3 demonstrates why it is considered good practice to perform thermodynamic measurements using **thermostatted** apparatus to ensure T does not vary, i.e. to eliminate scatter on graphs of p against V by obviating the need to consider more than one controlled variable.

....
A **thermostat** is a device for maintaining a temperature. The name comes from *thermo*, which is Greek for 'energy' or 'temperature', and 'stat' which derives from the Greek root *statikos*, meaning 'to stand', i.e. not move or alter.
....

Additional problems

7.1–5 Decide which of the following statements are qualitative and which are quantitative:

(a) I performed more experiments than you.

(b) The spectrum contains a peak at 500 nm.

(c) The lecture lasted 50 minutes.

(d) The lecture was longer than last week's lecture.

(e) Five hundred people attended the lecture.

7.6 Beer's law says the optical absorbance A of a chromophore is directly proportional to its concentration $[C]$. The data below refer to the absorbances *Abs* of aqueous solutions of permanganate ion MnO_4^- of concentration $[C]$. (The optical path length was 1 cm, and the wavelength of observation was 523 nm.)

$[C]$/mol dm⁻³	0	0.0001	0.0002	0.0003	0.0004	0.0005	0.0006
Absorbance, *Abs*	0	0.2334	0.4668	0.7002	0.9336	1.167	1.4004

Plot a graph of A (as y) against $[C]$ (as x).

7.7 When a mixture of volatile liquids is left to stand, the composition of the vapour above the liquid obeys **Raoult's law**: the amount of a material in the vapour mixture above the mixture relates to the vapour pressure of the *pure* liquid $p^\ominus$ and the mole fraction x of the component in the liquid mixture. These data relate to benzene.

Mole fraction, x	0	0.2	0.4	0.6	0.8	1
Partial pressure, p/Pa	0	14 940	29 880	44 820	59 760	74 700

Plot a graph of pressure p (as y) against mole fraction of benzene x (as x).

7.8 The ideal-gas equation suggests a correlation between the volume V of an ideal gas and its pressure p. Plot a graph of volume V (as y) against pressure p (as x) to ascertain whether such a relationship holds. The data refer to helium.

V/m^3	0.01	0.02	0.025	0.03	0.04	0.05	0.06	0.07	0.08	0.1	0.2
p/Pa	100	50	40	33.3	25	20	16.7	14.3	12.5	10	5

7.9 The ideal-gas equation suggests the volume of an ideal gas should be directly proportional to the temperature. Using the data below, plot a graph of volume V (as y) against temperature T (as x).

T/K	280	300	320	340	360	380	400	420	440
V/m^3	0.0233	0.0249	0.0266	0.0283	0.0299	0.0316	0.0333	0.0349	0.0366

7.10 Consider the data below, which relate to the second-order racemization of glucose in aqueous hydrochloric acid at 17 °C. The concentration of glucose is [A].

Time, t/s	0	600	1200	1800	2400
[A]/mol dm^{-3}	0.400	0.350	0.311	0.279	0.254

Plot a **second-order kinetic plot**, i.e. plot 1/[A] (as y) against time (as x).

8

Graphs II

The equation of a straight-line graph

The algebraic equation of a straight line is $y = mx + c$

By the end of this section, you will know:

- We can write the mathematical form of a straight-line graph as $y = mx + c$.
- We sometimes need to reduce the equation, if y is written with a factor.
- 'm' is the gradient of the graph.
- 'c' is the intercept of the graph on the y axis.
- We may need to *reduce* the equation if the equation has a form $n \times y = mx + c$, by dividing throughout by n.

The equation of a straight line will always take the algebraic form in eqn (8.1):

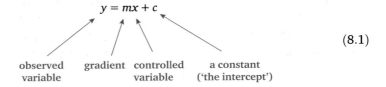

$$y = mx + c$$

| observed variable | gradient | controlled variable | a constant ('the intercept') |

(8.1)

> **Self-test 1**
>
> For each of the following, what is the gradient m and what is the intercept c?
>
> | | |
> |---|---|
> | **1.1** $y = 2x + 4$ | **1.4** $y = 4x + 3$ |
> | **1.2** $y = 3.5x - 2$ | **1.5** $y = x + 22$ |
> | **1.3** $y = 4x + 8.4$ | **1.6** $y = -4x - 4.3$ |

If we see an expression written to look like eqn (8.1), but with a factor in front of the y, in such cases it is common practice to divide throughout by that constant. We say we reduce the equation.

■ **Worked Example 8.1**

The relationship linking the electrochemical current I (in mA) through an electrode of area A (in cm^2) follows the form:

$$5.2I = 3A + 12.2$$

Reduce the relationship to the form $y = mx + c$.

In words, the equation indicates that for every three-fold increase in area, the current increases by a factor of 5.2. We deduce that current I is the observed variable y and the electrode area A is the controlled variable x. The additional current of 12.2 mA is constant, whatever the size of the electrode, and probably represents a fault in the circuitry measuring the currents.

To reduce the equation, we divide throughout by 5.2:

$$\frac{5.2}{5.2}I = \frac{3A}{5.2} + \frac{12.2}{5.2}$$

and, because $5.2 \div 5.2 = 1$, then:

$$I = \frac{3A}{5.2} + \frac{12.2}{5.2}$$

It would be silly to leave the equation like this, so we would then perform the two divisions, to yield:

$$I = 0.57A + 2.34$$

We might have written out the equation in this way:

$$\underset{y}{I} = \underset{m}{\left|\frac{3}{5.2}\right|} \underset{x}{|A|} + \underset{c}{\frac{12.2}{5.2}} \quad \blacksquare$$

Self-test 2

Reduce the following equations to the form $y = mx + c$, and determine the gradient m and the intercept c:

2.1	$2y = 4x + 10$	2.5	$30.2y = 19x + 10.4$
2.2	$3y = 5x - 30$	2.6	$-4y = -4x + 12$
2.3	$-5y = 10x + 15$	2.7	$3.17y = -2x + 1.22$
2.4	$7.1y = 4x$	2.8	$10^6y = 10^4x - 10^7$

'Satisfying' the equation of a line

By the end of this section, you will know that:

- Data satisfy the equation only if they lie on the straight line.
- A point not on the line does not satisfy the equation.

Consider a straight line of the form $y = 4x + 2$. If we know the value of x, then we can calculate the value of y. In this case, we say that y is a sum; in words, we say that y equals four times the value of x plus an additional two of $x = 0$, then $y = 2$. If $x = 1$, then $y = 6$, and so on. We could write this data as a table:

x	-3	-2	-1	0	1	2	3
$4x$	-12	-8	-4	0	4	8	12
$4x + 2 = y$	-10	-6	-2	2	6	10	14

We could then construct a graph in just the same way we drew Fig. 7.1, with the values of x from the table plotted against the respective values of y we just determined. It is convenient to group the data together: we often put them in brackets, with the value of x first, followed by a comma, then the value of y. In this example, we would write $(-3, 10)$, $(-2, -6)$, $(-1, -2)$, and so on. We might call these data pairs, or merely the points.

The point $(0,2)$ lies on this straight line, as does $(2, 10)$. We say these points satisfy the equation. The point $(2, 7)$ does not lie on the line, so we say it does *not* satisfy the equation. A moment's thought ought to persuade us that an infinite number of points will satisfy a single line, because a straight line can stretch from $x = -\infty$ to $+\infty$.

The intercept c

By the end of this section, you will know that:

- 'c' is the intercept of the graph on the y axis.
- The intercept on the y axis is only the same as 'c' if the x axis extends as far as $x = 0$.
- The intercept will be positive if it intercepts the y axis above the x axis and negative if it intercepts below the x axis.
- The intercept is zero if the line goes through the origin.
- A straight line occupies at least two of a graph's quadrants.

We can tell the value of the constant c at a glance, because the straight line strikes the y axis at a value of c. For this reason, it is quite common to see c called 'the intercept'. The intercept can have any value, and can be positive, negative, or goes through the origin, in which case the intercept is zero.

We do not see a constant c if the straight line goes through the origin of the graph; stated another way, the value of the constant c is 0 because the value of y at the origin is 0. Such a situation would arise if we had tared the weighing boat before weighing our sample on, say, a top-pan balance. If $c = 0$, we would then write either $y = mx + 0$ or even omit the constant, saying $y = mx$. These two equations mean the same.

If we have the equation of a line, we can read off the intercept as the final number, i.e. the figure expressed without an accompanying x or y. For example, if the equation is $y = 4x + 2$, then the intercept must be 2. It is *not* true, however, to say the intercept is the same as the constant if we forgot to first reduce the equation to the form $y = mx + c$, but had some multiple of y.

A **tare** is the allowance made for the weight of a weighing boat or the packaging encasing something. The law says we must tare the weight of the goods we buy in a shop to accommodate additional weights such as packaging. This explains why we see the words '*Net* weight' printed on a packet.

■ Worked Example 8.2

Having dissolved varying amounts of the organic dye crystal violet (**I**) in water, we determine the optical absorbance of each solution. Figure 8.1 is a Beer's law plot constructed with the data.   What is the figure's intercept?

I

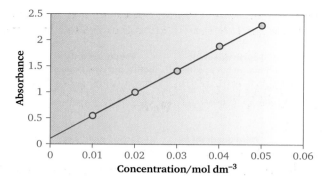

Figure 8.1 Graph of optical absorbance (as *y*) of Crystal Violet in aqueous solution against concentration of absorbing species in solution (as *x*): Beer's law.

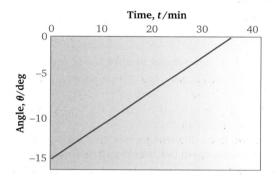

Figure 8.2 A beam of plane-polarized light twists through an angle of θ as it passes through a solution containing a chiral compound: graph of θ (as *y*) against time (as *x*) showing the way that θ decreases with time if racemization occurs during reaction.

The bold line touches the *y* axis at an absorbance of 0.1, so the intercept is 0.1. The equation of this straight line therefore has the algebraic form $y = mx + 0.1$. We have not been asked for the magnitude of the gradient *m*. ∎

Sometimes the intercept on a graph is negative.

> **Beer's law** suggests the line should pass through the origin. In this example, the cause of the graph's non-zero intercept indicates that we failed to 'blank' the spectrum.

■ **Worked Example 8.3**

A solution of the chiral alkyl halide (**II**) reacts with hydroxide ion to form a racemic mixture. When a beam of plane-polarized light is passed through the reaction mixture, its angle of twist θ decreases steadily with time until it reaches zero, according to the graph in Fig. 8.2.

What is the intercept?

II

The bold line in Fig. 8.2 intercepts both axes: it cuts the *x* axis at $t = 35$ min and crosses the *y* axis at $\theta = -15°$. We only need the intercept on the *y* axis when deriving the

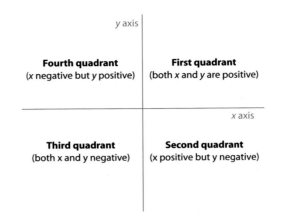

Figure 8.3 Graph labelled with the names of the four quadrants.

equation of the straight line, so we can ignore where the line crosses the x axis. Accordingly, the intercept is $-15°$.

We write the equation of this straight as either:

$$y = mx + (-15)$$

or:

$$y = mx - 15.$$

....
A **quadrant** of a Cartesian graph is one of the four segments produced by dividing a graph by both a horizontal and a vertical axis. The word comes from the Latin *quadrans* meaning a quarter.
....

The two equations mean the same, but we would usually use the latter style because it looks less complicated. ■

This last graph depicted data drawn *below* the x axis. In fact, we will need to plot data in any of the four quadrants of the axes. Figure 8.3 names the four quadrants. We will not employ this notation, but the names are so common that we will certainly encounter them in other books. A straight-line graph will always occupy at least two quadrants.

Occasionally, we find that a line has been expressed in the form $y = c + mx$ rather than the more familiar $y = mx + c$. Having learnt the algebra in Chapter 2, we should now recognize how this form is identical to the more familiar version, $y = mx + c$, except its order has been changed.

■ **Worked Example 8.4**

The ionic conductivity Λ of a solute in solution depends on its concentration $[C]$ according to the **Onsager equation**:

$$\Lambda = \Lambda^0 - b\sqrt{[C]}$$

....
Care: Do not confuse the 'C' in concentration $[C]$ with the intercept c, here.
....

where b is a constant, Λ is the ionic conductivity, and Λ^0 is the conductivity of the same compound at zero concentration.

How would we obtain Λ^0 graphically?

Methodology:

(i)   We first decide how the equation may be represented as a straight line.

(ii)   Next, we analyse the nature of the general equation $y = mx + c$, and allocate the terms m and c to the component parts of the Onsager equation.

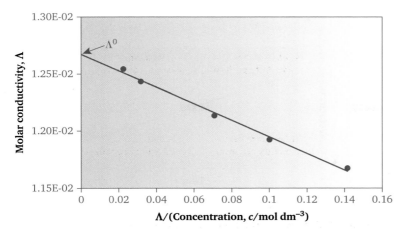

Figure 8.4 Onsager plot of molar conductivity of aqueous sodium nitrate (as *y*) against the square root of concentration (as *x*). The intercept of Λ° represents the molar conductivity at 'infinite dilution'.

The variables in this example are concentration [C] and conductivity Λ. (Strictly the Onsager equation tells us the controlled variable is the *square root* of concentration, $\sqrt{[C]}$.) Concentration [C] is the *controlled* variable here because we can readily prepare a series of solutions in the laboratory. Conductivity is the *observed* variable here because we do not measure a value of Λ until the solution of concentration [C] is available.

The term *b* is a constant; and Λ^0 is also a constant, because there can only be one value of conductivity at zero concentration. One of these constants will represent the intercept of a graph while the other will represent the gradient. To decide which is which, we note that if the Onsager equation is thought to describe a straight line, then the gradient will be the constant written adjacent to the controlled variable, and the intercept will be the constant that is not multiplied by any other term. Accordingly, *b* is the gradient and Λ^0 is the intercept. ∎

This choice of intercept makes physical sense, because it represents the conductivity at zero concentration, and the intercept on the *y axis* must of necessity also occur at zero concentration.

In summary:

$$\underset{y}{\Lambda} = \underset{c}{\Lambda^0} - \underset{m}{b} \underset{x}{\sqrt{[C]}}$$

Although we reversed the order of *c* and *mx* here, we have not altered the equation except in terms of the way we represent it. With equal validity, we could have written, $\Lambda = -b\sqrt{[C]} + \Lambda^0$, but writing a minus sign in front of the constant might confuse.

Figure 8.4 shows an Onsager plot of Λ (as *y*) against $\sqrt{[C]}$ (as *x*) for KNO_3 in water at 298 K. The intercept of Λ^0 is labelled.

We sometimes say Λ^0 is the conductivity at **infinite dilution**.

Self-test 3

Determine the intercept in each of the following graphs:

3.1 The velocity of an object decelerating: graph of velocity (as *y*) against time (as *x*).

» *continued* . . .

>> *Self-test 3 continued . . .*

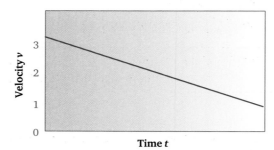

3.2   The amount of money earned: graph of salary (as y) against duration of the work (as x).

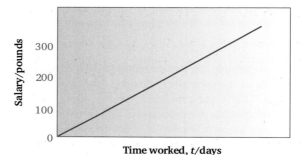

3.3   The current drawn through an electrochemical cell during electroplating: graph of current (as y) against the concentration of analyte (as x).

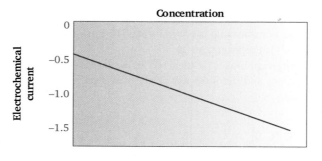

Quite often, we find we cannot just read off the intercept, because the x axis does not extend as far as $x = 0$. In this case, we will need to redraw the graph, or work out the equation of the graph and *calculate* the value of c.

■ **Worked Example 8.5**

The electrode potential of a redox couple is given the symbol $E_{O,R}$. If $E_{O,R}$ is determined under standard conditions, we call it the *standard* electrode potential, and give it the symbol $E_{O,R}^{\ominus}$. The two electrode potentials $E_{O,R}$ and $E_{O,R}^{\ominus}$ are related by the **Nernst equation**. The Nernst equation describes the couple comprising copper ion, copper metal as:

$$E_{Cu^{2+},Cu} = E_{Cu^{2+},Cu}^{\ominus} + \frac{RT}{2F}\ln[Cu^{2+}]$$

Determine the value of $E_{Cu^{2+},Cu}^{\ominus}$ from the data below:

Concentration $[Cu^{2+}]/mol\ dm^{-3}$	10^{-2}	10^{-3}	10^{-4}	10^{-5}	10^{-6}
$E_{Cu^{2+},Cu}/V$	0.281	0.251	0.222	0.192	0.162

We sometimes call the symbol ⊖ a **plimsoll sign**. For this reason, we sometimes say that $E_{O,R}^{\ominus}$ is 'E plimsoll'.

Methodology:

(i) As with Worked example 8.4 above, we must first decide which variable is the observed and which the controlled.

(ii) With these variables, we decide which of $E_{Cu^{2+},Cu}^{\ominus}$ and $(RT \div nF)$ is the gradient and which the intercept.

(iii) We plot a graph with these variables, and determine the intercept. ∎

In this example, the electrode potential $E_{Cu^{2+},Cu}$ is the observed variable (and will be plotted as *y*) and logarithm of concentration will be the controlled variable (and will be plotted as *x*). In other words, we 'cut up' the equation as follows:

$$\underset{y}{E_{Cu^{2+},Cu}} = \underset{c}{\left| E_{Cu^{2+},Cu}^{\ominus} \right.} + \underset{m}{\left. \frac{RT}{2F} \right|} \underset{x}{\ln[Cu^{2+}]}$$

Logarithms are outlined in Chapter 11.

From the arrangement of the equation, and with reasoning similar to that in Worked example 8.4, we see how $(RT \div nF)$ is the gradient and $E_{Cu^{2+},Cu}^{\ominus}$ the intercept. Figure 8.5 shows a Nernst plot of $E_{Cu^{2+},Cu}$ (as *y*) against the logarithm of $[Cu^{2+}]$ (as *x*) constructed using the data from the table.

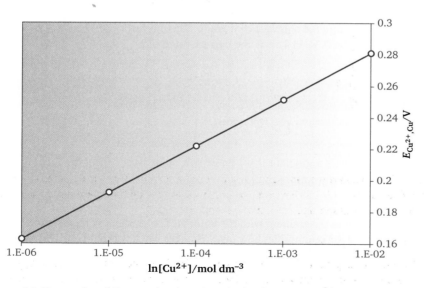

Figure 8.5 Nernst plot of $E_{Cu^{2+},Cu}$ (as *y*) against the logarithm of $[Cu^{2+}]$ (as *x*) to show the validity of the Nernst equation. The intercept is not the same as $E_{Cu^{2+},Cu}^{\ominus}$ because the *y* axis does not join the *x* axis at a value of 0.

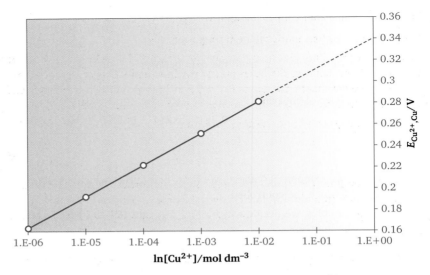

Figure 8.6 Nernst plot of $E_{Cu^{2+},Cu}$ (as y) against the logarithm of $[Cu^{2+}]$ (as x), constructed with the same data as in Fig. 8.5. In this figure, the intercept *is* the same as $E^{\ominus}_{Cu^{2+},Cu}$ because the y axis joins the x axis at a value of 0.

The figure is drawn in a naive way. In it, the line strikes the y *axis* at 0.281 V. The true intercept corresponds to $E^{\ominus}_{Cu^{2+},Cu}$, but we would be wrong to say that 'the intercept c is 0.281 V'. To prove this statement, we only need to substitute 0.281 V into the Nernst equation above as $E^{\ominus}_{Cu^{2+},Cu}$ and see how the data in the table do not fit the equation.

The reason why this intercept is not *the* intercept, i.e. $E^{\ominus}_{Cu^{2+},Cu}$, follows from the way we drew the x axis: a careful look at the scale beneath the x axis shows that it extends from 10^{-6} up to a value of 10^{-2}, i.e. does not extend as far as 0. In fact, to determine the *correct* value of $E^{\ominus}_{Cu^{2+},Cu}$, we need to replot the graph, but ensure the x axis extends as far as $x = 0$. Figure 8.6 shows such a graph. The intercept (i.e. the value of y when $x = 0$) is seen to be 0.34 V. This is the correct value of $E^{\ominus}_{Cu^{2+},Cu}$.

The gradient *m*

By the end of this section, you will know:

- How to determine the gradient of a graph by looking at it.
- How to determine a gradient if we know two points that satisfy the line.
- That a gradient will often have units.

We typically denote the gradient with the letter m. The gradient m represents a rate of change: in this case, it quantifies how fast y changes when we change x. The steepness of a hill is a simple example of such a gradient: when we see a sign describing the hill's gradient as '1 in 4', we should understand that the hill is steep, because it rises vertically by one unit for every four units we travel horizontally. A slope of '1 in 10' is gentler, and '1 in 100' is barely a hill at all.

We determine the gradient of a straight-line graph as follows:

1. We first draw the best straight line through our data.

2. At one end of the line, we choose a point on the line, and call it (x_1, y_1). We then choose a second point at the other end of the line, and call it (x_2, y_2). It does not matter where we choose the points (x_1, y_1) and (x_2, y_2), but in practice we tend to get a more accurate value if the two points are widely separated.

3. We define the gradient *m* according to eqn (8.2):

$$m = \frac{y_2 - y_1}{x_2 - x_1} \qquad (8.2)$$

where the top line (the **numerator**) represents the *vertical* distance travelled and the bottom line (the **denominator**) represents the *horizontal* distance.

4. If the *x* and *y* axes have units, then we retain them within the numerator and denominator in eqn (8.2).

Some people choose to write eqn (8.2) in a slightly different way, as in eqn (8.3):

$$m = \frac{\Delta y}{\Delta x} \qquad (8.3)$$

where the symbol Δ is an operator (see Chapter 1) which means 'change in'. We define Δ in eqn (8.4):

$$\Delta = \text{final form} - \text{initial form} \qquad (8.4)$$

> We often meet the operator Δ in thermodynamics, e.g. in ΔH, which we define as the difference between the final enthalpy and the initial enthalpy.

Equations (8.2) and (8.3) are identical, and only differ is terms of the way we cite them.

■ **Worked Example 8.6**

A man wishes to give up smoking. As part of his regime, he sticks a slow-release patch onto his arm. The patch releases the addictive alkaloid nicotine (**III**) at a precise and controlled rate. Before the observation period, the patch had released 0.25 millimoles of (**III**) and after a further three days, it had released a total of 0.4 millimoles. Determine the daily rate of nicotine release, i.e. calculate the gradient of a graph of amount of nicotine released (as *y*) against time elapsing in days (as *x*).

III

> *Remember*:
> The gradient of such a graph is a rate.

Before we start, we define terms. At the beginning, we say time $t = 0$, and at the end of the observation period $t = 3$ days. Similarly, at the beginning, $n_{(III)} = 0.25$ mol and $n_{(III)} = 0.4$ mol after 3 days.

Inserting values into eqn (8.2):

$$\text{rate} = \left(\frac{\text{change in observed variable}}{\text{change in controlled varibale}} \right) = \frac{0.4 - 0.25}{3 - 0} \frac{\text{mmol}}{\text{days}}$$

so:

$$\text{rate} = \frac{0.015\,\text{mmol}}{3\,\text{days}} = 0.015\,\text{mmol day}^{-1}$$

Sometimes the gradient m is negative. Such a gradient tells us that the value of y decreases as the value of x increases. ∎

■ Worked Example 8.7

During the course of a chemical reaction, the concentration of the reactant decreases from $0.01\,\text{mol dm}^{-3}$ at the start of the reaction to $0.03\,\text{mol dm}^{-3}$ after half an hour (30 minutes $= 1800\,\text{s}$).

What is the rate at which the reactant is consumed?

Inserting values into eqn (8.2):

$$\text{rate} = \frac{0.03 - 0.1\,\text{mol dm}^{-3}}{1800 - 0\,\text{s}}$$

so:

$$\text{rate} = \frac{-0.07\,\text{mol dm}^{-3}}{1800\,\text{s}}$$

and we calculate the rate as $-3.9 \times 10^{-5}\,(\text{mol dm}^{-3})\,\text{s}^{-1}$. ∎

> The minus sign of this rate (gradient) tells us the concentration *decreases* with time.

Self-test 4

Calculate the gradient m between the following points:

4.1	(1,2) and (2,4)	4.5	$(-3, -3)$ and $(-5, -5)$
4.2	(10,17) and (5,22)	4.6	$(2, -3)$ and (4,5)
4.3	(3,5) and (6,14)	4.7	$(-2, -8)$ and $(-4, -4)$
4.4	$(-2,3)$ and (0,7)	4.8	$(-5,4)$ and (7,12)

We often obtain a rate of change by measuring the magnitude of a gradient, rather than looking at two bits of data. In fact, a graph generally yields a superior gradient, because we have many data to choose from. We therefore minimize our errors. For example, it is usually clear which data are poor after plotting a graph. We ignore the bad data by drawing a straight line though only the good data. Determining a gradient from two bits of data is more risky: what if one or both were duff?

So the gradient of a graph yields directly a rate of change. For example, the gradient of a graph of velocity against time tells us how fast an object speeds up ('accelerates') with time.

■ Worked Example 8.8

The powerful magnets inside the chamber of a mass spectrometer cause the molecular ion of vitamin C (**IV**) to accelerate. Its velocity v increases steadily with time t as depicted in Fig. 8.7.

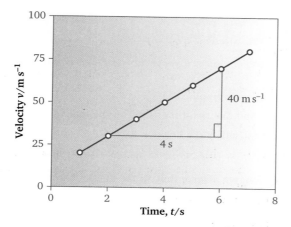

Figure 8.7 A straight-line graph to demonstrate how we determine the velocity of a particle travelling inside a mass spectrometer. The gradient yields the acceleration, *a*.

IV

The gradient of the graph directly yields the acceleration *a*. Determine from the graph (i) the ion's acceleration and (ii) the units of acceleration *a*.

(i) We first obtain a *numerical magnitude* for *a* using the *numbers* within eqn (8.3). To determine a value of the gradient, we first construct a right-angled triangle against the straight-line trace. By measuring along the horizontal and vertical limbs of the triangle, we quantify how the velocity *v* increases by $40\,\mathrm{m\ s^{-1}}$ during an interval of 4 s.

Inserting values from Fig. 8.7 into eqn (8.3), we calculate the acceleration:

$$a = \text{gradient} = \frac{40\,\mathrm{m\ s^{-1}}}{4\,\mathrm{s}}$$

so (**IV**) accelerates at $a = 10\,\mathrm{m\ s^{-2}}$.

(ii) Then we derive the *units* of *a* from the arrangement of the *units* within the fraction. In this case, *a* has the units of the numerator divided by the units of the denominator:

units of $a = \mathrm{m\ s^{-1}} \div \mathrm{s^{-1}}$

In words, we say the velocity increases by 10 metres per second, for every second the molecule resides within the spectrometer. We manipulate these units algebraically to obtain $\mathrm{m\ s^{-2}}$.

By looking carefully at the units of a gradient, we often possess a powerful method of working out a parameter. For example, a graph of distance covered (as *y*) against time (as *x*): the gradient has the units of 'distance ÷ the units of time', i.e. $\mathrm{m\ s^{-1}}$, which helps us realize that the gradient of the graph represents velocity. ∎

Determining the equation of a straight line

By the end of this section, you will know how to determine the equation of a straight line:

- By looking carefully at a graph.
- From the gradient and one point that satisfies the line.
- From two points that satisfy the line.

As we start this section, we assume that all the lines are straight, so each follows the form $y = mx + c$. In practice, it is generally easier to start with the intercept since we merely read it from the y axis. We need to be careful that we are reading the *real* intercept, though, because the intercept is the value of y when $x = 0$: if the x axis does not go as far as zero, then we are looking at *an* intercept rather than *the* intercept, which is the value of c in the equation. If the graph is drawn in such a way, we may have to redraw it with a longer x *axis* that does go as far as $x = 0$.

■ **Worked Example 8.9**

The graph in Fig. 8.8 describes the temperature inside a calorimeter (as y) as compound is consumed (as x). Deduce the equation of the straight line.

Firstly, the solid line in Fig. 8.8 strikes the y axis at a value of $y = 10$, so the constant c has a value of 10. An incomplete equation for the line is therefore $y = mx + 10$.

Next, we obtain the gradient m: we note how the line passes through the two points $(0, 10)$ and $(15, 40)$. We call these points respectively (x_1, y_1) and (x_2, y_2), and insert the respective values into eqn (8.2), which yields:

$$m = \frac{40 - 10}{15 - 0}$$

so:

$$m = \frac{30}{15} = 2$$

We deduce the equation of the line to be $y = 2x + 10$. ■

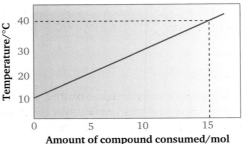

Figure 8.8 Energy is released when a compound is burnt within a bomb calorimeter. This energy causes the temperature inside the calorimeter to rise: graph of calorimeter temperature (as y) against amount of compound burnt (as x).

Self-test 5

Determine the equation of the straight line in each figure. Take care with the signs.

5.1

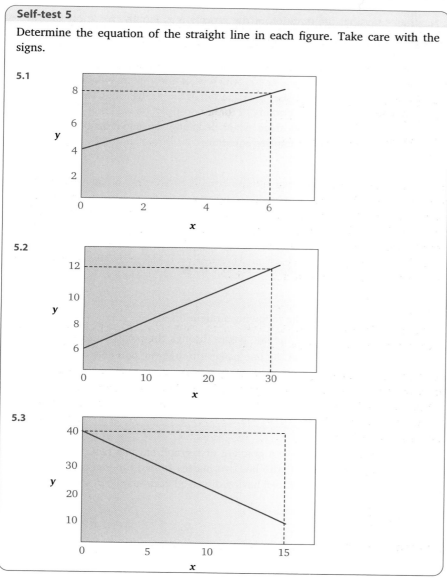

5.2

5.3

Sometimes we don't have a graph, so we use the data directly. Most commonly, we possess the gradient of the straight line and know the coordinates of a single point that lies on it.

■ **Worked Example 8.10**

A factory produces sugar at a rate of 4 tonnes per hour. After 1 hour, the warehouse contains 12 tonnes of sugar. Write an equation to relate the amount of sugar in the warehouse and the process time, assuming none of the sugar is removed.

The amount of sugar is measured in tonnes, so any value of y has the units 'tonnes'.

We start by defining terms. We note there are two variables to think about: the amount of sugar produced, and the length of time the process runs for. Let's say the amount of sugar in the warehouse is y. We'll call this quantity the observed variable.

The length of the process x is measured in hours, so any value of x has the units 'hours'.

Next, let's say that the duration of the process is x. Finally, we know the **rate** of sugar production is 4 tonnes per hour, so we equate the rate with the gradient m. While the numerical value of m is 4, it is a rate with units of 'tonnes per hour'.

The equation of a straight line is $y = mx + c$. Having equated the rate of production with the gradient m, we say $y = 4x + c$, which is an 'incomplete equation'.

We know one point that satisfies the equation of the straight line: the question says, 'After 1 hour, the warehouse contains 12 tonnes of sugar', so we know the point $x = 1$, $y = 12$. Therefore, the second stage when deducing the equation of the line is to substitute values into the incomplete equation:

$$12 = 4 \times 1 + c$$

so:

$$12 = 4 + c$$

and:

$$c = 8$$

The equation of the line is therefore $y = 4x + 8$. ∎

Self-test 6

Deduce the equations of the following straight lines:

6.1   The gradient is 3 and the line passes through the point (1, 2).

6.2   The gradient is 10 and the line passes through the point (4, 4).

6.3   The gradient is -4 and the line passes through the point (3, 0).

6.4   The gradient is 2.5 and the line passes through the point (4, 5).

6.5   The gradient is -3.5 and the line passes through the point (-4, 0).

At other times, we do not have a gradient or a graph but just two pairs of data. An obvious example is a chemist who has two masses from a top-pan balance and subsequently two optical absorbances, having dissolved each sample in solvent and taken the spectrum.

■ Worked Example 8.11

An analytical chemist wants to prepare a calibration graph, relating the amount of the natural pigment β-Carotene (**V**) with its optical absorbance when in solution. The analyst dissolves 0.01 g of (**V**) and obtains an optical absorbance of 0.8, then weighs a mass of 0.03 g and obtains a higher absorbance of 2.0. What is the relationship between the absorbance (the observed variable, y) and the mass of (**V**) (the controlled variable, x)?

V

It's usually a good idea to restate the data, as (0.01, 0.8) and (0.03, 2.0). We first determine the gradient m by inserting the data into eqn (8.2)

$$m = \frac{2.0 - 0.8}{0.03 - 0.01g}$$

so:

$$m = \frac{1.2}{0.02\,g}$$

and:

$$m = 60\,g^{-1}$$

The incomplete equation is therefore $y = 60\,x + c$.

We then insert one of the data pairs into this equation to obtain the value of c. It does not matter which point we insert. Indeed, it is often a good idea to perform the calculation twice, in order to verify our answer. We will use the second point, (0.01,0.8):

$$0.8 = 60 \times 0.01 + c$$

so:

$$0.8 = 0.6 + c$$

and:

$$c = 0.2$$

We deduce the equation of the line is $y = 60x + 0.2$.

Chemists often find it useful to rephrase such a relationship in words, saying something like, 'absorbance $= (60 \times \text{mass}) + 0.2$'. ∎

Self-test 7

Deduce the equations of the straight lines connecting the following pairs of points:

7.1 (1,2) and (2,4)

7.2 (0,− 2) and (3,− 11)

7.3 (9,12) and (26,50)

7.4 (7,9) and (6,10)

7.5 (1,2) and (− 3,− 4)

7.6 (− 3,− 3) and (− 2,− 6)

Additional problems

8.1 The amount of chemical converted per hour is 400 kg. The amount of the chemical in the store at the start of the day is 312 kg. Assuming none of the chemical is removed during the course of the day, derive an equation to describe the amount of chemical in the store as the day progresses.

8.2 A lecture course is so easy that the exam marks M depend only on attendance A rather than ability. For every lecture attended, the exam mark increases by 1.5%. Prior knowledge means that a student attending nothing still gains a mark of 18%. Deduce an equation relating exam mark M and attendance A.

8.3 'Reduce' the following equation: $12y = 4x - 6$.

8.4 What is the equation of a straight line of gradient 4.2, which passes through the point (3, 1)?

8.5   The temperature voltage coefficient of an electrode or cell is a *rate* of change:

$$\text{temperature voltage coefficient} = \frac{E_{\text{(cell)}2} - E_{\text{(cell)}1}}{T_2 - T_1}$$

Calculate the value of this rate of change (gradient) for the Clarke cell, if the value of $E_{\text{(cell)}}$ increases from 1.433 V to 1.456 V as the temperature is lowered from 56 °C to 25 °C.

8.6   Deduce the equation of the straight line having a gradient of -4.9, and which passes through the point $(-5,2)$.

8.7   A hill is steep. For every 20 metres forward, it goes up 3 metres. Calculate its gradient, m.

8.8   In a modified form of the Beer–Lambert equation, the optical absorbance *Abs* and concentration $[C]$ of chromophore in solution as follows:

$$Abs = \varepsilon[C]l + k$$

where ε is the extinction coefficient, l is the path length, and k is the absorbance of the glass cell. Deduce the value of k if the absorbance *Abs* was 0.8, the concentration $[C]$ was 0.23 mol dm^{-3}, the path length l was 1 cm, and the extinction coefficient was 30 mol^{-1} dm^3 cm^{-1}.

....
We use this version of the **Beer–Lambert law** when we cannot blank the cell before measuring the absorbance of the sample.
....

....
In reality, a pH meter is merely a pre-calibrated voltmeter.
....

8.9   A chemist notes the relationship between the pH of a solution and the reading on the voltmeter, E. When the pH is 4.2, the reading is 300 mV, and when the pH is 5.5, the reading is 400 mV. Deduce the form of the values of m and c in the equation of a line of the form $E = \text{pH} \times m + c$.

8.10   The ideal-gas equation $pV = nRT$ suggests a correlation between the volume V of an ideal gas and its pressure p. Plot a graph of volume V (as y) against the reciprocal of pressure p (as x) and ascertain the equation of the straight line. The data refer to carbon dioxide.

V/m^3	0.01	0.02	0.025	0.03	0.04	0.05	0.06	0.07	0.08	0.1	0.2
p/Pa	50	25	20	16.5	12.5	10	8.3	7.2	6.2	5	2.5

Graphs III

Straight lines that intersect

When straight lines intersect

By the end of this section, you will know that:

- Two straight lines are parallel if they have the same gradient but different intercepts.
- Two straight lines will meet and cross over, unless they are parallel.

Two straight lines will be parallel if their gradients are identical but their intercepts differ. For example, $y = 4x + 5$ and $y = 4x + 10$ in Fig. 9.1 are parallel, and separated vertically on the page by five units. Many calibration curves appear as parallel lines.

With the sole exception of parallel lines, a pair of straight lines will always intersect eventually. When we first looked at straight lines, we considered the equation $y = 4x + 2$. We saw then how such an equation is satisfied by an infinite number of points, because the values of x can vary from $-\infty$ through to $+\infty$. It's a long line, however small the portion we choose to draw or think about. In fact, a moment's thought should persuade us that all straight lines are infinitely long, again because there are an infinite number of values that x can take.

But now think about two lines that are not parallel. For example, $y = 4x + 2$ and $y = 6x + 6$. The simplest definition of parallel lines says, 'lines that never meet'. By contrast, lines that are not parallel do meet: we say they **cross over**, or **intersect**.

> The simplest definition of **parallel lines** says, 'lines that never meet'.

While each line is satisfied by an infinite number of points, only a single point can simultaneously satisfy two equations. This should be obvious: the two lines cross only when the value of x is the same for both and the value of y is the same for both. At no other values of x and y will a single point satisfy both equations. The point is unique.

It is often crucial to know at what values of x and y two straight lines meet, and the point of intersection often represents crucial physicochemical parameters.

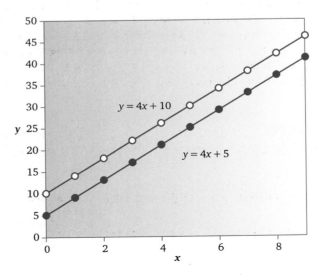

Figure 9.1 Parallel lines never meet: the two lines $y = 4x + 5$ and $y = 4x + 10$ are parallel because they have the same gradients m but different intercepts c.

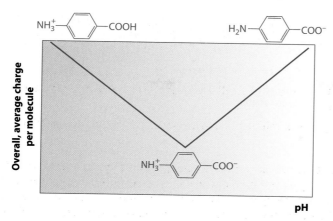

Figure 9.2 Two separate trends appear when we consider the overall, average charge per molecule: anions form at high pH and cations at low pH. The two straight lines *intersect* at the so-called **isoelectric point**, when a proportion of **(I)** exists as a **zwitterion**, i.e. it bears both a positive and a negative charge.

■ **Worked Example 9.1**

Consider the molecule *p*-aminobenzoic acid, **(I)**. The molecule is always ionized: in acidic solution, the amine portion is protonated and in alkaline solution the acid dissociates to form a carboxylate ion.

H₂N —⟨ ⟩— COOH

I

Draw a schematic graph of the overall charge on **(I)** as a function of pH. ■

> **Schematic** means 'stylized' or 'general'.

There are many simple ways to determine the isoelectric point. One of the simplest is to obtain data about the ionic charges at high pH and at low pH, i.e. at the extremities of the graph. We then draw straight lines through the data, and extend them both towards the centre of the graph: we say we **extrapolate** the lines. The isoelectric point represents the pH where the two **extrapolants** intersect on Fig. 9.2.

A simple way to obtain the charge data is from measurements of ionic conductivity Λ.

> The **isoelectric** point represents the pH when the molecule **(I)** exists as a zwitterion, bearing both a positive and a negative charge.

Graphical solutions of simultaneous equations

By the end of this section, you will know:

- That the point of intersection is a single point with a unique value of *x* and *y*.
- How to determine the point of intersection by drawing a graph.

This example of an isoelectric point suggests the simplest way to determine the point of intersection is to draw two lines on a single graph, and observe where they cross.

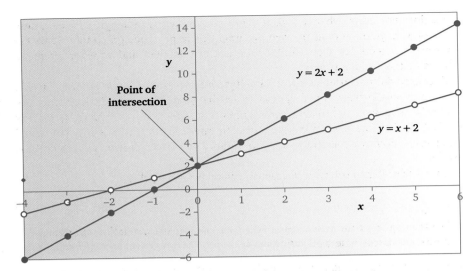

Figure 9.3 The two straight lines $y = 2x + 2$ and $y = x + 2$ intersect at the single, unique point (0,2).

■ **Worked Example 9.2**

Consider the two lines $y = x + 2$ and $y = 2x + 2$. By drawing both lines on a single graph, determine the single point of intersection.

Figure 9.3 shows both lines drawn on the same graph, and clearly shows the unique point of intersection is (0,2).

Because we draw both straight lines at the same time, we call this method the use of **simultaneous equations**. ■

Self-test 1

By plotting a graph in each case, determine the point of intersection between the following pairs of straight lines:

1.1 $y = x + 2$ and $y = 4x + 5$

1.2 $y = 6x - 12$ and $y = -6x + 12$

Algebraic solution of simultaneous equations

By the end of this section, you will know:

- Graphical methods of determining the point of intersection are not efficient, unless x and y are whole numbers.
- How to determine the point of intersection using the algebraic methods of elimination and substitution, and by multiplying.

Graphical methods generally yield the point at which two lines intersect. The method is a little time consuming, though, and prone to significant errors if the intersection point does not occur when x and/or y are not whole. For example, if $x=4.1$ and $y=3.44$, it might require us to draw two graphs: a large, crude graph that shows the intersection occurs 'at about (4,3),' then a more detailed and enlarged graph to determine the decimal places.

Chemists do not usually solve simultaneous equations by drawing a graph. They may be too busy, but they also know that graphical methods are inherently imprecise. It may be easy to read from a graph an intersection point like $(1, 2)$, but it is more difficult if the point does not represent integers, say, $(1.23, 2.04)$. We would need very good graph-plotting skills to obtain the correct answer to two decimal places.

Chemists prefer to solve simultaneous equations with algebra.

....
An **integer** is a whole
number.
....

■ **Worked Example 9.3**

We look again at the two lines $y=x+2$ and $y=2x+2$: determine the point of intersection, this time using algebra.

We first write the two equations one above the other in the form of a simple sum:

$$\begin{aligned} y &= 2x+2 \\ -y &= \ x+2 \end{aligned}$$

We then **eliminate** the y terms by subtracting one equation from the other. To do this, we subtract one term at a time, so we say '$y-y=0$', '$2x-x=x$', and '$2-2=0$'. Therefore, by elimination we obtain:

$$\begin{aligned} y &= 2x+2 \\ -y &= \ x+2 \\ \hline 0 &= \ x+0 \end{aligned}$$

So the value of x which satisfies the two equations is $x=0$. If the value of x is 0 when the two lines intersect, and the point also lies on both lines, then we can obtain the value of y at the point of intersection by **substituting** $x=0$ into either of the two equations that intersect.

Taking the equation $y=2x+2$ and inserting $x=0$, we obtain $y=0+2$, so the value of y at the point of intersection is 2. The point of intersection is $(0,2)$. We obtain the same result when substituting into the other equation, $y=x+2$. ■

In the Worked example immediately above, we eliminated the y terms because the two values of y were the same. In practice, we subtracted one equation from the other. We can employ an identical approach if the two values of x are the same.

■ **Worked Example 9.4**

Use algebra to determine the point of intersection between the two lines $y=x+2$ and $3y=x+6$.

As before, we start by writing the two equations one above the other, and eliminate the x terms by subtracting:

$$\begin{aligned} 3y &= x+6 \\ -y &= x+2 \\ \hline 2y &= 0+4 \end{aligned}$$

so $2y = 4$ and so $y = 2$. We determine the value of x at the point of intersection by substituting $y = 2$ into either of the two equations:

$$3 \times 2 = x + 6 \quad \text{so } x = 0$$

$$\text{or} \quad 2 \quad = x + 2 \quad \text{so } x = 0$$

so the two straight lines intersect at $(0,2)$. ∎

By substituting into both equations and obtaining the same value of x, we corroborate that the value of y is common to both, i.e. is indeed y at the point of intersection.

Self-test 2

Algebraically determine the point of intersection between the following pairs of straight lines:

2.1 $y = 2x + 5$ and $y = 11 - 4x$

2.2 $y = x + 2$ and $y = -3x - 14$

2.3 $y = 4x + 7$ and $y = 2x + 5$

2.4 $2y = x + 10$ and $2y = 6x + 10$

2.5 $y = 2x + 3$ and $y = 4x + 3$

2.6 $5y = 4x + 2$ and $y = 4x + 6$

Occasionally, we employ a variation of this method, because the factors of x or y are negative.

■ **Worked Example 9.5**

Determine the point of intersection between the two lines $y = -x + 2$ and $2y = x + 4$ using algebra.

As before, we start by writing the two equations one above the other, but because one x term is positive and the other negative, this time we *add* the two equations:

$$2y = x + 4$$
$$+y = -x + 2$$
$$\overline{3y = 0 + 6}$$

If the factors on x are negative in one equation and positive in the other, we *add* the two equations to ensure the two x terms cancel.

So $3y = 6$ and so $y = 2$. Substituting this value of y into either of the two starting equations yields $x = 0$. The point of intersection is $(0,2)$. ∎

Self-test 3

Algebraically determine the point of intersection between the following pairs of straight lines:

3.1 $y = 2x + 5$ and $-y = -4x + 11$

3.2 $y = x + 2$ and $y = -3x - 14$

3.3 $2x + y = 4$ and $2x + 2y = 6$

3.4 $x + y = 7$ and $3x + 2y = 17$

3.5 $x - y = -1$ and $x + 2y = 8$

3.6 $2x + 2y = 32$ and $3x - 2y = 3$

3.7 $x + y = 16$ and $5x + y = 60$

3.8 $2x - y = 2$ and $x + y = 5.5$

Sometimes we can formulate a pair of equations from a physical problem.

■ **Worked Example 9.6**

Find two numbers x and y such that their sum is 12 and their difference is 4.

We first formulate a pair of simultaneous equations from the data in the question. Firstly, the sum of the numbers is 12, so $x + y = 12$. Secondly, their difference is 4, so $x - y = 4$.

We can add the two equations in order to eliminate y:

$$
\begin{aligned}
 x + y &= 12 \\
 +x - y &= 4 \\
 \hline
 2x &= 16
\end{aligned}
$$

so $x = 8$. Clearly, substitution reveals that $y = 4$.

The equations we've just looked at are simple because we can eliminate either the value of x or the value of y by writing one equation above the other. This level of simplicity is rare. We usually need to manipulate the data a little before we can eliminate. ■

■ **Worked Example 9.7**

A mass spectroscopy experiment suggests that two molecular fragments contain only carbon and hydrogen. The empirical formula of the first is C_2H_6 and has a mass of 30 Da and the second has a formula of C_6H_{13} and a mass of 85 Da.

What are the masses of carbon and hydrogen atoms?

We first re-express the data, saying $2C + 6H = 30$ and $6C + 13H = 85$. These equations only *look* more complicated than most of the equations above because we wrote H and C rather than x and y.

We could write the two sets of data in the form of algebraic equations, as follows:

$$2C + 6H = 30 \quad (1)$$
$$6C + 13H = 85 \quad (2)$$

> The SI unit of atomic mass is the **dalton**, which has the symbol **Da**.

Clearly, no simple algebraic solution is possible yet, but if we multiply (1) by 3, we obtain the pair:

$$6C + 18H = 90 \quad 3 \times (1)$$
$$6C + 13H = 85 \quad (2)$$

Subtracting (2) from '3 × (1)' yields:

$$
\begin{aligned}
 6C + 18H &= 90 \quad 3 \times (1) \\
 -6C + 13H &= 85 \quad (2) \\
 \hline
 0 \quad 5H &= 5
\end{aligned}
$$

so:

$$5H = 5 \text{ Da and } H = 1 \text{ Da}$$

One hydrogen atom has a mass of 1 Da. Knowing the mass of one hydrogen atom, we then calculate the mass of carbon by substituting $H = 1$ into either equation to yield the result, $C = 12$. ■

> This example might yield data we knew already, but the calculation in this worked example demonstrates the power of the method.

■ **Worked Example 9.8**

According to the Beer–Lambert law, the absorbance of a single **chromophore** in solution is A and relates to the concentration $[C]$, the optical path length l, and the extinction coefficient ε:

$$A = [C]\varepsilon l$$

> A **chromophore** is a molecule of species that imparts a colour, such as a dye.

A solution contains two chromophores, A and B. The extinction coefficient of dye A is ε_A, and ε_B relates to dye B. The path length l remains constant at 0.1 m, but the concentrations of A and B vary. The absorbance is 1.0 when $[C]_A = 0.2 \, \text{mol dm}^{-3}$ and $[C]_B = 0.4 \, \text{mol dm}^{-3}$; the absorbance drops to 0.7 when the concentration of A is $0.1 \, \text{mol dm}^{-3}$ and $[C]_B = 0.3 \, \text{mol dm}^{-3}$. Use simultaneous equations to determine values of ε_A and ε_B.

....

A **spectrometer** is a device for measuring optical absorbances, A.

....

Before we start, we note how a spectrometer measures the sum of the absorbance of A and B: $A_{\text{measured}} = A_A + A_B$. From the Beer–Lambert law, we can substitute for the two absorbances, yielding:

$$A_{\text{measured}} = [C]_A \varepsilon_A l + [C]_B \varepsilon_B l$$

Next, we rewrite the data from the question concerning A, $[C]$, and l terms, to obtain a pair of simultaneous equations:

$$1.0 = 0.02 \times \varepsilon_A + 0.04\varepsilon_B \qquad (1)$$
$$0.7 = 0.01 \times \varepsilon_A + 0.03\varepsilon_B \qquad (2)$$

These simultaneous equations are similar in form to those above. The first difference is cosmetic: we wrote ε_A and ε_B rather than x and y. Otherwise they are similar; in fact, such a difference is unimportant. A second and more serious difference is an inability to eliminate either ε_A or ε_B because their coefficients differ. Any algebraic solution would be easier if the coefficients on either ε_A or ε_B were the same.

But the coefficients can be the same if we multiply eqn (2) by a factor of 2:

$$1.4 = 0.02 \times \varepsilon_A + 0.06\varepsilon_B \qquad 2 \times (2)$$
$$-1.0 = 0.02 \times \varepsilon_A + 0.04\varepsilon_B \qquad (1)$$

We then subtract (1) from $2 \times (2)$ to eliminate ε_A, in much the same way as earlier in Worked examples 9.5 and 9.8:

$$1.4 = 0.02 \times \varepsilon_A + 0.06\varepsilon_B \qquad 2 \times (2)$$
$$\underline{-1.0 = 0.02 \times \varepsilon_A + 0.04\varepsilon_B} \qquad (1)$$
$$0.4 = 0 \times \varepsilon_A \quad + 0.02\varepsilon_B$$

so:

$$0.4 = 0.02 \times \varepsilon_B$$

and:

$$\varepsilon_B = \frac{0.4}{0.02} = 20$$

We deduce by straightforward substitution into either (1) or (2) that $\varepsilon_A = 10$. ■

It is sometimes quite difficult to decide what the factors to use.

■ **Worked Example 9.9**

The liquids benzene (**II**) and bromobenzene (**III**) are mixed. The vapour above this mixture contains both (**II**) and (**III**). The pressure above pure benzene is $p_{(\text{II})}^{\ominus}$ and the pressure above pure bromobenzene is $p_{(\text{III})}^{\ominus}$.

According to Raoult's law, the pressure of the gases above this mixture p_{total} is given by the equation:

$$p_{\text{total}} = p_{(\text{II})}^{\ominus} \times x_{(\text{II})} + p_{(\text{III})}^{\ominus} \times x_{(\text{III})}$$

where x is the proportion in the liquid, expressed as a fraction. The value of p_{total} is 72 kPa when $x_{(\text{II})} = 0.3$ and $x_{(\text{III})} = 0.7$, and 80 kPa when $x_{(\text{II})}$ and $x_{(\text{III})}$ are both 0.5.

Use simultaneous equations to determine values for $p_{(II)}^{\ominus}$ and $p_{(III)}^{\ominus}$.

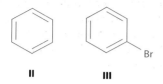

II   III

The value of $p_i^{\ominus}$ is called the **saturated vapour pressure** of i. The higher value of $p^{\ominus}$ for benzene means it is more volatile than bromobenzene.

We first rewrite the data from the question in the form of two equations:

$$72 = 0.3p_{(II)}^{\ominus} + 0.7p_{(III)}^{\ominus} \qquad (1)$$

$$80 = 0.5p_{(II)}^{\ominus} + 0.5p_{(III)}^{\ominus} \qquad (2)$$

It is not possible to solve these two equations without additional factors. A moment's study suggests we cannot simply multiply one of the equations by a factor: we will need to multiply both by additional factors; and the two factors will need to be different.

In a case like this:

(i) We first decide which variable we wish to eliminate. In this example, we will eliminate the two $p_{(II)}^{\ominus}$ terms. Each value of $p_{(II)}^{\ominus}$ already has a coefficient: in this case, $p_{(II)}^{\ominus}$ has been multiplied by 0.3 in eqn (1) and by 0.5 in eqn (2). We need to modify the two coefficients such that they become the same so the $p_{(II)}^{\ominus}$ term in one equation can be subtracted completely from the other.

(ii) To make them the same, we multiply the coefficient in eqn (1) by the factor in eqn (2), and multiply the coefficient in eqn (2) by the factor in eqn (1). (In fact, we often employ multiples of the factors.)

To ensure the two factors for $p_{(II)}^{\ominus}$ are the same, we multiply eqn (1) by 5 and eqn (2) by 3:

$$360 = 1.5p_{(II)}^{\ominus} + 3.5p_{(III)}^{\ominus} \qquad 5 \times (1)$$

$$240 = 1.5p_{(II)}^{\ominus} + 1.5p_{(III)}^{\ominus} \qquad 3 \times (2)$$

We can now subtract the lower equation from the upper:

$$360 = 1.5p_{(II)}^{\ominus} + 3.5p_{(III)}^{\ominus} \qquad 5 \times (1)$$

$$\underline{-240 = 1.5p_{(II)}^{\ominus} + 1.5p_{(III)}^{\ominus}} \qquad 3 \times (2)$$

$$120 = 0 \qquad + 2.0p_{(III)}^{\ominus}$$

In this example, we multiply eqn (1) by 5, and multiply eqn (2) by 3, rather than 0.5 and 0.3 respectively, because the final numbers are easier to manipulate.

so $2.0p_{(III)}^{\ominus} = 120$, and $p_{(III)}^{\ominus} \div 2 = 60$ kPa.

Substitution into eqn (1) or eqn (2) yields $p_{(II)}^{\ominus} = 100$ kPa and $p_{(III)}^{\ominus} = 60$ kPa. ∎

Self-test 4

Algebraically determine the point of intersection between the following pairs of straight lines:

4.1 $y = -2x + 10$ and $y = x + 1$

4.2 $y = 2x + 1$ and $y = -3x + 11$

4.3 $2x = -3y + 48$ and $3x = -2y + 37$

4.4 $y = x - 3$ and $-4y = 5x - 6$

4.5 $5x + 2y = 4.5$ and $3x + 4y = 5.5$

4.6 $0.5y = 2.5x + 7$ and $0.25y = -5x + 1$

Additional problems

9.1–5 Which of the following pairs of lines are parallel?

(i) $y = 4x - 2$ and $y = 3x - 2$

(ii) $7.1y = 9.2x + 5$ and $7.1y = 9.2x - 6.23$

(iii) $y = 5x + 1$ and $2y = 10x$

(iv) $0.2y = x + 11$ and $4y = 20x + 2$

(v) $y = x + 1$ and $11.1y = 11.3x$

9.6 The mass of a sample in a weighing boat is 12 g. The mass of a second batch comprising twice the amount of sample, but in the same weighing boat, is 23 g. Derive an equation relating amount of sample and hence determine the mass of the boat.

9.7 Using a suitable graphical method, determine the point of intersection between the following straight lines: $y = 7x + 2$ and $y = 4x - 3$.

9.8 The Onsager equation relates ionic conductivity Λ and concentration $[C]$:

$$\Lambda = \Lambda^0 - b\sqrt{[C]}$$

where b is a constant and Λ^0 is the conductivity at 'infinite dilution'. Two solutions are prepared: the first has a concentration $[C]$ of $1.1 \times 10^{-4}\,\mathrm{mol\,dm^{-3}}$ and a conductivity Λ of $0.016\,\mathrm{S\,m^{-1}}$, and the second has a concentration of $1.58 \times 10^{-4}\,\mathrm{mol\,dm^{-3}}$ and a conductivity of $0.02\,\mathrm{S\,m^{-1}}$. Determine values for Λ^0 and b.

9.9 The Gibbs function G is defined by the expression $G = H - TS$, and the enthalpy is given by $H = U + pV$. Eliminate H from the first expression. [Hint: first rearrange the second expression to make U the subject.]

9.10 A pH meter is in reality a pre-calibrated voltmeter. The relationship between pH and the *emf* of the voltmeter is given by the expression:

$$emf = K + 0.059\ \mathrm{V} \times \mathrm{pH}$$

A pH electrode is immersed in a solution of buffer at pH = 7.0, and the *emf* measured as 276 mV. What is the pH when the pH electrode is subsequently immersed in a solution and the *emf* is 502 mV? [Hint: convert all *emf* values to volts before starting.]

10

Powers I

Introducing indices and powers

Introducing powers: the *base* and *index* of a number

By the end of this section, you will know:

- Basic nomenclature relating to base, power, index, and exponent.
- The phrase a^b means the number a has been multiplied by itself b times.

Power terms dominate the algebra we encounter in chemistry. Not only do they appear in a large number of our equations, but we need to know how to manipulate them in order to use the equations effectively.

At heart, power terms can be reduced to units of the type a^b.

Concerning relationships of the type a^b:

- We call a the base.
- We call b the index of the base, or its power.
- We sometimes see b called the exponent.
- In words, we would say, 'a raised to the power of b', 'a to the power of b', or just 'a to the b'. Each means the same.
- We always write the exponent as a ^{super}script, and place it to the *right* of the base, so a^b is correct but a_b or $^b a$ is not.
- Anything expressed with a power of 1 is the thing itself.

■ **Worked Example 10.1**

Consider cyclobutadiene (I) and cubane (II). The length of each C–C bond is l. Write an expression for the area of (I) and the volume of (II) in terms of l. For each expression, identify the base and the index.

I   II

Assuming compound (I) is square, we write its area A as $l \times l = l^2$. Here, l is the base and the index is 2.

The volume of (II) $V = l \times l \times l = l^3$. Here, the base is again l but this time the index is 3.

These examples illustrate the simplicity of indices. A few more examples:

- The expression 4^5 means 4 multiplied by itself five times: $4 \times 4 \times 4 \times 4 \times 4$.
- b^3 means b multiplied by itself three times, i.e. $b \times b \times b$.
- In general, we can write b^c, which means b multiplied by itself c times.

For simplicity, we sometimes omit the multiplication signs. We can write *cccc* instead of c^4; and *XXXXX* means the same as X^5. We can write the algebraic expression for the volume of a sphere as:

$$\text{either} \left(\frac{4 \times \pi \times r \times r \times r}{3}\right) \text{ or } \left(\frac{4}{3}\pi r^3\right)$$

Most people thinks the latter looks tidier, but the two versions of the expression are identical. ∎

Sometimes, manipulating exponents can greatly simplify an expression.

∎ Worked Example 10.2

A chemical plant processes 10 000 litres of solution per hour. We wish to increase the rate a hundred-fold. Calculate the new processing rate.

We can answer this worked example using two different approaches:

- firstly, using simple arithmetic; or
- secondly, by manipulating the indices.

Simple algebra

The new flow rate is $10\ 000 \times 100$ litres hour^{-1} $= 1\ 000\ 000$ litres hour^{-1}, so simple arithmetic says we process 1 million litres of solution per hour.

Indices

We first write each number as a power, so 10 000 becomes $10 \times 10 \times 10 \times 10$, and 100 becomes 10×10. We then simply add up the number of instances where 10 appears. We obtain:

$$\text{new rate} = (10 \times 10 \times 10 \times 10) \times (10 \times 10) = 10^6$$

Instances of 10   1   2   3   4   5   6

The sum involves six instances of 10, so the answer is 10^{six} (i.e. 10^6, 1 million).

An index of 1

Usually, we do not write the index against the base if its value is 1, because anything with an index of 1 is the thing itself. For example, we would write length $= l$ rather than l^1, although the latter is perhaps more useful when trying to understand the mathematics of indices. ∎

We need to know:

- We generally do not write the index against the base if its value is 1.
- Anything with an index of 1 is the thing itself.

Examples:

7 to the power of 1 (i.e. 7^1) has a value of 7.

Further examples:

$24^1 = 24$ $a^1 = 1 \times a$ $b^1 = 1 \times b$ $x^1 = 1 \times x.$

Multiple terms:

Mathematical phrases incorporating powers can be joined together.

■ Worked Example 10.3

Express $c \times c \times b \times b \times b$ in terms of bases and indices.

Methodology:

(i)   We first collect the like terms together.

(ii) We next determine the magnitude of the indices.

(i)   Collecting together the like terms: $c \times c \times b \times b \times b = (c \times c) \times (b \times b \times b)$.

(ii) Rewriting in terms of indices: $(c \times c) \times (b \times b \times b) = c^2 b^3$. ■

Negative indices

By the end of this section, you will know that:

• A negative index indicates a fraction.

• The larger the number after the index sign, the smaller the number.

■ Worked Example 10.4

A solution of concentrated acid has a concentration of $1.0 \, \text{mol} \, \text{dm}^{-3}$. We dilute it 10-fold. Calculate the new concentration of the acid, expressing the answer in the form of 10 to the power of an index.

Simple inspection alone tells us the new concentration is $0.1 \, \text{mol} \, \text{dm}^{-3}$, which we obtain as a simple fraction:

$$\frac{1 \, \text{mol} \, \text{dm}^{-3}}{10} = 0.1 \, \text{mol} \, \text{dm}^{-3}$$

In index form, we call this answer 10^{-1}. ∎

We need to know:

- The 10 merely tells us that we are working in base 10.
- The minus sign tells us that a fraction is involved.
- The index of 1 in this example tells us that the bottom line of the fraction contains one instance of 10.

This concentration is expressed incorrectly until we write the units after the number.

■ **Worked Example 10.5**

An analyst prepares a standard solution containing $1 \, \text{mol} \, \text{dm}^{-3}$ of sodium, starting with a solution containing $1 \, \text{mol} \, \text{dm}^{-3}$. The analyst progressively dilutes the solution: 10-fold, 10-fold again, and then a further 10-fold.
   What are the concentrations of sodium in the three solutions?

The concentration after the **first** dilution is $\frac{1}{10} \, \text{mol} \, \text{dm}^{-3} = 10^{-1} \, \text{mol} \, \text{dm}^{-3}$.
The concentration after the **second** dilution is $\frac{1}{10 \times 10} \, \text{mol} \, \text{dm}^{-3} = 10^{-2} \, \text{mol} \, \text{dm}^{-3}$.
The concentration after the **third** dilution is $\frac{1}{10 \times 10 \times 10} \, \text{mol} \, \text{dm}^{-3} = 10^{-3} \, \text{mol} \, \text{dm}^{-3}$.
We see how larger negative indices point towards smaller concentrations. ∎

Self-test 3

Perform the following divisions, expressing the answer as a base and index:

3.1 $1 \div a^3$

3.2 $1 \div c^2$

3.3 $1 \div (b \times b \times b)$

3.4 $1 \div (dddd)$

3.5 $1 \div z^{3.3}$

3.6 $1 \div (p^{4.1} \times p^{9.2})$

3.7 $1 \div (d \times d^4)$

3.8 $1 \div (jjjjkkkk^2hhh)$

Self-test 4

Write an expression for each of the following, and leave your answer as a fraction:

4.1 2^{-1}

4.2 10^{-2}

4.3 100^{-2}

4.4 6^{-3}

4.5 5^{-5}

4.6 13^{-3}

A special case: the index of 0

By the end of this section, you will know:

- Anything expressed with a power of 0 has a value of 1.

When we write area $A = l^2$, in effect we mean $A = 1 \times l \times l$. Similarly, the volume $V = l^3$ means $V = 1 \times l \times l \times l$. It is general practice to ignore the initial part which says '$1 \times \cdots$' but we need to remember that it is there.

■ **Worked Example 10.6**

A solution of sulphuric acid has a concentration of 10.0 mol dm^{-3}. We progressively dilute the solution four times, first 10-fold, then 10-fold again, and so on. Express the concentrations of the four acid solutions in index form.

(i)   Initially, the concentration is 1×10 mol dm^{-3}.

(ii)   The concentration after a 10-fold dilution is $(1 \times 10 \div 10) = 1$ mol dm^{-3}.

(iii) A further 10-fold dilution yields a concentration of $(1 \div 10) = 0.1$ mol dm^{-3}.

(iv)   Yet another 10-fold dilution yields a concentration of $(0.1 \div 10) = 0.01$ mol dm^{-3}. ■

In index form, the concentration of solution (i) is 10^1 mol dm^{-3}, the concentration of solution (iii) is 10^{-1} mol dm^{-3}, and solution (iv) has a concentration of 10^{-2} mol dm^{-3}.

From this trend, we see how the concentration of solution (ii) must be 10^0 mol dm^{-3} when written in index form. This is the only logical result. In fact, we derive the enormously important result: anything written to the power of 0 has a value of 1.

....
Anything expressed with an index of **0** has a value of 1.
....

Examples:

$2^0 = 1$ $3^0 = 1$ $4000^0 = 1$ $(a\,b\,c)^0 = 1$ anything$^0 = 1$.

The trend is illustrated for powers of 10 in the range million to millionth in Table 10.1.

Exploring the algebra of indices

By the end of this section, you will know:

- When **multiplying** terms expressed in terms of bases and indices, $a^x \times a^y = a^{(x+y)}$.
- When **dividing** terms expressed in terms of bases and indices, $a^x \div a^y = a^{(x-y)}$.
- When we have anything a raised to a power x, that is itself raised to a further power y, then the two powers are multiplied: $(a^x)^y = a^{(x \times y)}$.

■ **Worked Example 10.7**

A litre flask contains 1000 cm^3 of solution. What is the volume (in cm^3) contained within a hundred litre flasks?

As before, we can perform the calculation in two ways: by intuition and using bases and indices.

Intuition

Simple arithmetic tells us the total volume is 'volume per flask $\times$ number of flasks', so the volume is 10^3 cm$^3 \times 100 = 100\,000$ cm^3.

Table 10.1 Powers of 10 from 10^6 to 10^{-6}, written in words, value and in index form.

Name	Value	10^n
million	10 000 000	10^6
hundred thousand	100 000	10^5
ten thousand	10 000	10^4
thousand	1000	10^3
hundred	100	10^2
ten	10	10^1
one	**1**	$\mathbf{10^0}$
tenth	0.1	10^{-1}
hundredth	0.01	10^{-2}
thousandth	0.001	10^{-3}
ten thousandth	0.0001	10^{-4}
hundred thousandth	0.000 01	10^{-5}
millionth	0.000 001	10^{-6}

Bases and indices

We first convert each number to its appropriate index form:

- The volume of a single flask is $1000\,\text{cm}^3$, i.e. $10^3\,\text{cm}^3$.
- There are 100 flasks, i.e. 10^2.
- Because $100\ 000 = 10 \times 10 \times 10 \times 10 \times 10 \times 10$, the final volume is $10^5\,\text{cm}^3$.

We should note the pattern emerging:

$$10^3 \times 10^2 = 10^5$$

Looking closely at the indices, we see how $3 + 2 = 5$, so:

$$10^3 \times 10^2 = 10^5 \quad \text{is the same as} \quad 10^{(3+2)} = 10^5$$

If we multiply together two numbers, and each has been expressed in terms of 10 to the power of an index, then we can express the answer as 10 raised to a *new* index. We obtain this new index as the *sum* of the constituent indices. Note how we have, in effect, turned a multiplication problem into an addition problem. We can write the generalization for numbers expressed in base 10:

$$10^x \times 10^y = 10^{(x+y)} \tag{10.1}$$

In fact, we can write eqn (10.1) in a more general form still for any base:

$$a^x \times a^y = a^{(x+y)} \tag{10.2}$$

where a can be any number—indeed, the value of a need not be a whole number (an **integer**) but could be fractional. ■

Self-test 5

Rewrite the following multiplication problems, expressing each in the form a^b:

5.1 $10^2 \times 10^4$ 5.6 $7^2 \times 7^4$

5.2 $10^3 \times 10^5$ 5.7 $b^9 \times b^2$

5.3 $10^0 \times 10^2$ 5.8 $z^{15} \times z^2$

5.4 $10^{20} \times 10^{40}$ 5.9 $b^{4.1} \times b^{7.2} \times b^{3.8}$

5.5 $6^3 \times 6^{12}$ 5.10 $k^{6.22} \times k^{8.12}$

Working with exponents this way represents a very powerful way of simplifying.

■ Worked Example 10.8

Consider the unimolecular dissociation of ethane (III) to form two methyl radicals:

III

The reaction proceeds in the gas phase with the following rate law:

rate $= k \times [C_2H_6]$

where k is the rate constant, and has a value of 5.36×10^{-4} at 700°C.
What is the rate when $[C_2H_6] = 10^{-2}$ Pa?

Pa here is **pascal**, the SI unit of pressure.

Inserting numbers into the equation:

rate $= 5.36 \times 10^{-4} \times 10^{-2}$

so:

rate $= 5.36 \times 10^{[(-4)+(-2)]}$

i.e. rate $= 5.36 \times 10^{-6}$. ■

To simplify this calculation, we have omitted all the units.

■ Worked Example 10.9

A bottle of reagent holds a mass of 1000 g. We want to weigh out one-hundredth of this amount. What mass do we want?

To solve the problem longhand, we write:

$$\frac{1000\,g}{100} = 10\,g$$

Again, we could have written the sum in terms of indices:

$$\frac{10^3 g}{10^2} = 10\,g$$

We could have obtained this same result by noting how the term on the bottom could have been written as a^{-y}. In effect, then, $a^x \div a^y$ is the same as $a^x \times a^{-y}$. According to eqn (10.2), this *multiplication* sum yields $a^{(x-y)}$.

Once more, we notice a pattern emerging, with the index of the answer coming from the *difference* between the indices within the fraction. We write the general expression:

$$\frac{10^x}{10^y} = 10^{(x-y)} \tag{10.3}$$

This expression shows once more the power of this method of working. We can also write the algebraic expression in a yet more general way:

$$\frac{a^x}{a^y} = a^{(x-y)} \tag{10.4}$$

We can manipulate numbers in just the same way as we manipulate variables: for example, we can write $1 \div 2$ as 2^{-1}. Again, 10^{-1} is the same as $1 \div 10$. ■

Self-test 6

Use eqn (10.4) to express each of the following division problems in the form a^b:

6.1	$10^2 \div 10^4$	6.6	$1.5^2 \div 1.5^4$
6.2	$10^3 \div 10^5$	6.7	$b^9 \div b^{-2}$
6.3	$10^0 \div 10^{12}$	6.8	$z^5 \div z^{2.5}$
6.4	$10^{-2} \div 10^4$	6.9	$z^{1.3} \div z^{2.7}$
6.5	$4^3 \div 4^{12}$	6.10	$c^{3.15} \div c^{2.93}$

■ **Worked Example 10.10**

How many cubic centimetres are contained within a cubic metre?

As before, we can solve this problem in two ways: a simple, intuitive approach, and using the laws of powers.

Intuitive approach

The length 1 m contains 100 cm, therefore a cubic metre contains

$$(1\,\text{m})^3 = (100\,\text{cm})^3$$

so:

$$1\,\text{m}^3 = 1\,000\,000\,\text{cm}^3$$

A cubic metre contains a million cubic centimetres.

Approach using the laws of powers

$1\,\text{m} = 10^2\,\text{cm}$, so $(1\,\text{m})^3 = (10^2\,\text{cm})^3$ and:

$$(1\,\text{m})^3 = 10^2\,\text{cm} \times 10^2\,\text{cm} \times 10^2\,\text{cm} = 10^6\,\text{cm} \times \text{cm} \times \text{cm}$$

so:

$$1\,\text{m}^3 = 10^6\,\text{cm}^3$$

If we look closely, we see a new pattern emerging: when we have a number raised to a power x, that is itself raised to a new power y, then we multiply the two powers:

$$(10^2)^3 = 10^6$$

Or, in general:

$$(a^x)^y = a^{(x \times y)} \tag{10.5}$$

■

> The phrase **cubic metre**, i.e. $(1\,\text{m})^3$, means the volume enclosed by the space $1\,\text{m} \times 1\,\text{m} \times 1\,\text{m}$; and a **cubic centimetre** is the volume $1\,\text{cm} \times 1\,\text{cm} \times 1\,\text{cm}$.

> We can manipulate units in much the same way as we manipulate numbers and letters.

Self-test 7

Use eqn (10.5) to express each of the following problems in the form a^b:

7.1 $(10^2)^5$ 7.4 $(a^3)^7$

7.2 $(10^3)^{10}$ 7.5 $(p^{4.4})^{1.2}$

7.3 $(a^7)^3$ 7.6 $(7^{3.3})^{7.8}$

Fractional indices

By the end of this section, you will know that:

- A fractional index of $\frac{1}{2}$ implies a square root.
- A fractional index of $\frac{1}{3}$ implies a cube root.
- In general, a fractional index of $1/n$ implies the nth root.

We have already encountered a few *fractional* indices above. It is now time to understand what they actually *mean*, physically. We start by looking at square roots.

> ### ■ Worked Example 10.11
>
> Rewrite $\sqrt{16}$ in terms of a base and index.

In effect we are asking the question, 'What number do we need to multiply by itself to yield 16?' Clearly, the answer is 4.

But we could have rephrased the question somewhat. In a square-root problem, the 'something to be squared' is 16 raised to some as yet unknown power, which we will call x: we write 16^x instead of $\sqrt{16}$. Our problem here is to discover the value of x.

If 16^x is the square root of 16, then 16^x multiplied by itself will equal 16. And in terms of bases and indices, the number 16 can be written as 16^1.

In summary:

$$(16^x)^2 = 16^1$$

Using the relationship in eqn (10.5):

$$16^{2x} = 16^1$$

If the two halves of this equation are indeed the same, then the powers on either side must be the same, so:

$$2x = 1$$

which can only be true if the value of x is ½. Therefore, we indicate a square root by writing a power of ½. ■

A few simple examples:

$$9^{1/2} = 3 \qquad 144^{1/2} = 12 \qquad 100^{1/2} = 10.$$

With similar reasoning, a power of a third $\frac{1}{3}$ indicates a **cube root**. The two expressions $\sqrt[3]{27}$ or $27^{1/3}$ are equally valid.

In general, we indicate the nth root of a number or phrase by writing the power $1/n$. A few examples are:

- The fourth root of $p = p^{1/4}$.
- The third root of $8 = 8^{1/3}$.
- The yth root of $x = x^{1/y}$.

> ### Self-test 8
>
> Express each of the following word problems in the form $a^{1/n}$:
>
> 8.1   The third root of 27　　　　8.4   The ninth root of $(7b)$
>
> 8.2   The square root of 16　　　8.5   The jth root of 12
>
> 8.3   The fourth root of p　　　　8.6   The ith root of k

Additional problems

10.1 The volume of a sphere is given by the equation:

$$V = \frac{4}{3}pr^3$$

Rearrange this expression to make r the subject, expressing the answer in terms of indices.

10.2 The current I induced in a conductor when a voltage V is applied across a resistor R is given by Ohm's law:

$$V = IR$$

Rearrange the expression to make I the subject, and calculate the magnitude of the current I when a voltage V of 100 V is applied across a resistance R of $10^{12}\,\Omega$.

10.3 Dynamite is a high explosive. The reason why dynamite is so effective is because it produces a large volume of gas in a short period of time. If $1\,dm^3$ contains $1000\,cm^3$, and the volume of gas produced is $100\,dm^3$, re-express the volume produced in cm^3.

10.4 An electrochemist immerses a rotated-disc electrode, RDE in a solution of analyte at a concentration of [C]. The viscosity of the solution is v. The RDE rotates at a frequency of ω. The current at the RDE is given by the **Levich equation**:

$$I = 0.62\,nFA\,[C]\,D^{\frac{1}{2}}\,\omega^{\frac{1}{2}}\,v^{\frac{1}{6}}$$

where A is the area of the disc, D is the speed at which the analyte diffuses through solution, n is the number of electrons transferred, and F the Faraday constant. Rearrange this equation to make viscosity v the subject of the equation.

10.5 Take the Levich equation from Additional problem 10.4 above, and rewrite it, changing each index for the appropriate root sign.

10.6 The relationship between the energy of a photon, the speed of light c, the Planck constant h, and the wavelength of light λ is:

$$E = h^1 c^1 \lambda^{-1}$$

Rewrite this expression using a more conventional notation.

10.7 The definition of concentration [C] is:

$$[C] = \frac{amount}{volume}$$

Staring from this definition, show that the IUPAC units of concentration should be $mol\,dm^{-3}$.

10.8 Two particles experience a force when they approach, the magnitude of which varies with their separation r according to the **Lennard–Jones equation**:

$$\text{force} = 4\varepsilon\left\{\left(\frac{1}{r}\right)^{12} - b\left(\frac{1}{r}\right)^{6}\right\}$$

where b is a constant and ε is the permittivity. Multiply out the brackets, and then rewrite in terms of indices.

10.9 Chemical bonds can vibrate, much like two solid balls separated by a spring. The wavenumber of the vibration v is a function of force constant of the bond k and the reduced mass μ:

$$v = \frac{1}{2\pi c}\sqrt{\frac{k}{\mu}}$$

c is the speed of light. Rewrite the right-hand side as a single line rather than a fraction, i.e. using the appropriate indices.

10.10 Charged species experience an electrostatic interaction ϕ when they approach, the magnitude of which is given by **Coloumb's law:**

$$f = \frac{z^+ z^-}{4\pi\varepsilon_0\varepsilon\, r^2}$$

where r is the interparticle separation, z^+ and z^- are the respective charges, and ε is the permittivity. Rewrite the right-hand side as a single line rather than a fraction, i.e. using the appropriate indices.

11

Powers II

Functions of exponentials and logarithms

Introducing exponential functions

By the end of this section, you will know:

- An exponential function is generally written as e^x, but is occasionally written as exp(x).
- The rate of change of an exponential function is extremely rapid.

In everyday language, we often talk about 'exponential' increases like 'prices are increasing exponentially'. In using this word, we imply an extremely fast rate of increase. Examples in everyday life of the exponential function include:

- 'Price inflation' (i.e. the rate at which shop prices increase).
- The mass of a bacterium as it grows in a pond or flesh wound.

Examples in chemistry include:

- The rate of a chemical reaction increasing as the temperature increases.
- The rate of a first-order reaction decreasing with time since the reaction started.
- The amount of radionucleotide decreasing with time.
- The number of neutrons formed during a nuclear chain reaction, e.g. within an atomic bomb based on uranium-235.

Mathematically, one of the simplest **exponential** equations is $y = a^x$, where a is a constant, x is the controlled variable, and y is the observed variable.

> ■ **Worked Example 11.1**
> During nuclear decay, a neutron hitting an atom of uranium-235 causes it to split.
> On splitting, each atom of uranium-235 emits three more neutrons; see Fig. 11.1.
>   Show how the number of neutrons increases exponentially.

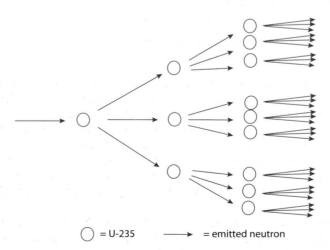

$\bigcirc$ = U-235 $\longrightarrow$ = emitted neutron

Figure 11.1 Schematic diagram to show how the number of uranium-235 atoms decaying during nuclear fission increases with the proportions $1 \to 3 \to 27 \to 81 \to 243 \to \ldots$.

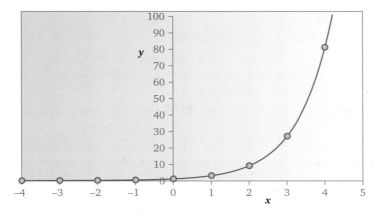

Figure 11.2 Graph of the exponential function $y = 3^x$.

Each atom of uranium-235 that splits ejects three neutrons, each of which causes a further atom of uranium-235 to split, so a further three atoms split during the 'second phase' of nuclear disintegration. Again, each of these atoms emits three neutrons, so a total of nine neutrons are formed, and so on.

The number of atoms splitting therefore increase in the series $1 \to 3 \to 27 \to 81 \to 243 \to \dots$. In fact, the numbers of splitting atoms obey the simple function $y = 3^x$. In the sequence above, $x = 0$, 1, 2, 3, then 4. If x had a value of 1/2, then the value of $y = 3^x$ would be 1.732. If x was -3, then $y = 1 \div (3^3)$, so $y = 1/27 = 0.037$.

Figure 11.2 shows a graph of this function, which clearly increases rapidly. In fact, it illustrates a central feature of exponentials: the rate at which the gradient of the graph changes continually increases in proportion to the index x on 3. In other words, the graph becomes steeper at a rapidly increasing rate. ■

We need to notice on this graph:

- The graph passes through the point (0,1), because any variable expressed to the power of 0 has a value of 1.

- The values of y are small when x is negative.

- The values of y are large when x is positive.

- The graph eventually increases at a very fast rate.

While we have used the word 'exponential' of this graph, in fact the word is generally used of a special type of function. In nature, we encounter many fundamental constants. The most common is pi (π), the ratio of a circle's circumference to its diameter. In Chapter 6, we met tau (τ), a constant that occurs often in nature. The third natural constant is e, and is in many respects the most commonly encountered.

Like the natural constants pi and tau, the value of e is *irrational*, which means we cannot express its value completely by writing it as a decimal number. To four significant figures, its value is 2.718. It might be more meaningful to write '2.718...'.

The simplest exponential function is $y = (2.718\dots)^x$, although we generally write it as e^x. We write 'e' here merely as a shorthand notation. This notation style is intended to remove the need to write '2.178...' each time.

We need to know:

- The exponential function 'e' is an operator.
- We say that x is the **domain** of the exponential operator, where x could be a number (whole or fractional), or an algebraic function of any sort.
- When we refer to e^x, we say aloud, 'e to the x'.
- Some people prefer to write the exponential operator as 'exp' instead of 'e': $y = e^x$ is the same as $y = \exp(x)$. The two styles are equally correct.
- We generally write 'exp' rather than 'e' to mean exponential when the domain is quite large, since the expression is less confusing and easier to read.
- Writing 'exp' instead of 'e' is often a good idea in chemistry, because 'e' can mean other things to a chemist:
  - an electron in electrochemistry;
  - (occasionally) an electronic charge;
  - in organic chemistry, an electrophile.

■ Worked Example 11.2

The rate of a chemical reaction k depends exponentially on the temperature T, according to the **Arrhenius equation**:

$$k = A \exp\left(-\frac{E_a}{RT}\right)$$

where R, E_a, and A are all constants.

Give the mathematical name of each part of the equation.

> Notice how we write 'exp' here rather than the shorter 'e', because the domain (the part of the equation enclosed within the brackets) is large and rather cumbersome.

- The operator is an exponential (written here as 'exp').
- The rate constant k on the left-hand side is the **observed variable**.
- The temperature T is the **controlled variable**.
- The portion of the equation within the brackets is the **domain**.
- The constant A is a **factor**. ■

> We often call the term A the **pre-exponential factor**: 'pre' because it is written *before* the exponential operator, and 'factor' because the exponential has been multiplied by it.

■ Worked Example 11.3

What is the exponential of 2.5?

In the past, determining the value of the exponential $e^{2.5}$ entailed a time-consuming calculation, requiring a slide rule or a book of tables—we call such a book 'a logarithm table' or (for short) 'log tables'. We can now employ a pocket calculator or personal computer.

With a calculator

> On most calculators, the e^x key is often positioned directly above the $\ln$ key. We may have to access the e^x function by pressing the second funtion or Inverse function key.

On most pocket calculators, the exponential key has the symbol .

(i)   Type in 2.5.

(ii) Press the e^x key.

The display will say 12.18 (ignore all the other, superfluous, significant figures).

On some calculators, steps (i) and (ii) are reversed.

With a PC

This example assumes we wish to use the Excel™ spreadsheet program.

(i) Place the cursor in the cell in which the calculation is wanted.

(ii)   Type '$=$EXP(2.5)'. (Note: The letters 'EXP' here are not case sensitive.)

(iii) Press the RETURN key. The display will probably say, 12.18249. ∎

In fact, the display may contain many more significant figures, the exact answer depending on the cell width. As professional chemists, we would not want to be seen citing a number to a silly number of significant figures. To obtain a *sensible* number of significant figures, we next:

(i)   Select the FORMAT drop-down menu.

(ii)   Select the CELLS page. The top of the box will say 'FORMAT CELLS'.

(iii) Select the tab saying NUMBER.

(iv)   Scroll down to NUMBER. A small box will allow you to choose the number of decimal places. Choose '3', for example, and press RETURN.

∎ **Worked Example 11.4**

One form of the van't Hoff isotherm is:

$$K = \exp\left(-\frac{\Delta G^{\ominus}}{RT}\right)$$

where K is an equilibrium constant, $\Delta G^{\ominus}$ the corresponding value of the standard change in the Gibbs function, R the gas constant, and T the absolute temperature
   Determine the value of K when $\Delta G^{\ominus} = 4$ kJ mol^{-1}, $R = 8.314$ J K^{-1} mol^{-1}, and $T = 298$ K.

Before inserting values, we must remember to accommodate the standard factor k, which is shorthand for '$\times 1000$'. Inserting values into this expression yields:

$$K = \exp\left(-\frac{4000 \, \text{J mol}^{-1}}{8.314 \, \text{J K}^{-1}\text{mol}^{-1} \times 298 \, \text{K}}\right)$$

We inserted 4000 rather than just 4 on the top line because the question tells us that ΔG° has a value of 4 **kilo**joules per mole; and the standard factor 'kilo' means '$\times 1000$'.

Incidentally, notice how all the units here have cancelled. This observation illustrates an important truth: we cannot obtain an exponential or a logarithm of anything except a number. So:

$$K = \exp\left(-\frac{4000}{2478}\right) = \exp(-1.614)$$

   i.e. $K = 0.199$. ∎

Self-test 1

Determine a value for each of the following exponentials:

1.1 e^3

1.2 $e^{3.2}$

1.3 e^{14}

1.4 $e^{-3.2}$

1.5   Repeat the calculation in Worked example 11.4, but use the different value $\Delta G^{\circ} = -12$ kJ mol^{-1}.

1.6   Repeat the calculation in Worked example 11.4, but use the different value $\Delta G^{\circ} = -40.2$ kJ mol^{-1}.

Exponential graphs

By the end of this section, you will know:

- An exponential graph increases rapidly when the domain is positive—we call it growth.
- An exponential graph decreases rapidly when the domain is negative—we call it decay.

Exponential graphs follow two general shapes, called 'growth' or 'decay', which are shown schematically in Fig. 11.3. The crucial difference between the two graphs is the sign of the domain.

So, when a graph is exponential, the gradient of the line will change rapidly:

- If the domain is *positive*, the graph gets *steeper* exponentially.
- If the domain is negative, then the graph gets *shallower* exponentially.

The inverse of exponentials: logarithm operators, and the algebra of their interconversion with exponentials

By the end of this section, you will know:

- The inverse function to exponential (e^x) is natural logarithms ($\ln x$).
- 10^x and logarithms expressed in base 10 are also algebraic inverses.

It is always necessary to be able to reverse an operation in order to obtain the domain x on its own. To reverse an exponential function e^x, we need to consider its inverse, which is a logarithm.

We give the symbol $\ln$ to a logarithm, where the 'l' stands for 'logarithm' and the 'n' stands for 'natural'. The logarithm of a domain x is written as: $\ln x$.

The letters $\ln$ stands for 'logarithm, natural'. The word 'natural' implies base 'e'. 'Unnatural' logarithms are expressed in a different base than 'e', and are introduced on page 134, below.

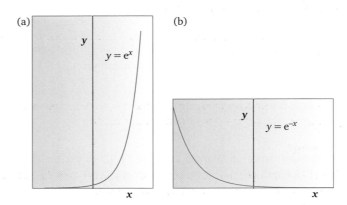

Figure 11.3 Schematic representations of exponential graphs (a) 'growth' with a positive domain, and (b) 'decay' with a negative domain.

It is easy to convert between e^x and its inverse $\ln x$ since they represent inverse functions:

$$e^x = y \quad \text{so} \quad \ln y = x \qquad (11.1)$$

Since the exponential and the logarithm are inverse functions, the domain x is common to both.

■ **Worked Example 11.5**

Rearrange the van't Hoff isotherm from Worked example 11.4 to make $\Delta G^{\ominus}$ the subject.

If we had MULTIPLIED a number by (for example) 4 and then performed the inverse function and DIVIDED by 4, we would have obtained the original number. In the same way, by taking the inverse of a function, we obtain the domain on its own.

The inverse function to exponential is natural logarithm 'ln', so to rearrange the equation, we must first take the logarithm of both sides:

$$\ln\left[\exp\left(-\frac{\Delta G^{\ominus}}{RT}\right)\right] = \ln K$$

As the inverse of an exponential is a logarithm, so $\ln(e^x)$ is just 'x'. In this way, we can simplify the left-hand side:

$$-\frac{\Delta G^{\ominus}}{RT} = \ln K$$

Finally, because the $\Delta G^{\ominus}$ term on the left-hand side has been divided by $-RT$, we multiply both sides by this same factor:

$$\Delta G^{\ominus} = -RT \ln K \quad ■$$

■ **Worked Example 11.6**

At the start of a reaction, the concentration of reactant is $[R]_0$. During the course of the reaction, which occurs via first-order kinetics, the concentration of R decreases with time according to a so-called **rate equation**:

$$[R]_t = [R]_0 e^{-kt}$$

Here, $[R]_t$ is the concentration of reactant R at times t after the reaction started, and k is a first-order rate constant.

Rearrange this equation to make $-kt$ the subject.

We will follow the BODMAS rules learnt previously from Chapters 3–5. Firstly, we divide both sides of the equation by $[R]_0$:

$$\frac{[R]_t}{[R]_0} = e^{-kt}$$

Next, because $-kt$ is the domain of an exponential, we perform the inverse operation to exponential. In words, we say 'take the natural logarithm':

$$\ln\left(\frac{[R]_t}{[R]_0}\right) = \ln\left(e^{-kt}\right)$$

But a function of its own inverse represents the original function. In other words, the right-hand side of the equation simplifies, and we obtain:

$$-kt = \ln_e\left(\frac{[R]_t}{[R]_0}\right) \quad ■$$

Care: ln ('ell en') is often (incorrectly) written with the first letter as i or I (eye) or even as a numeral **1** (one).

We occasionally see the operator 'ln' written as '$\log_e$' to emphasize how it depends on e. We will avoid this notation style here: ln automatically means logarithm in base e.

Logarithms in bases other than e

By the end of this section, you will know:

- We can express numbers to other **bases** than e, writing a^x, where a is a constant.
- The most common base other than e is base 10.
- We express each logarithm in terms of a **base** and a **domain**.
- The inverse to the function 10^x is again a logarithm.
- Logarithms expressed in base 10 are written as '$\log x$'.

So far, we have talked about logarithms as being the inverse function to the exponential 'e', where $e = 2.178\ldots$. In fact, we could take many other numbers, and write a similar equation, e.g. $y = a^x$, where a is a constant. We say that a is the **base** and x is the **domain**.

The only common example of equations of this sort is $y = 10^x$, i.e. where the base is 10. Just as a logarithm 'ln' is the inverse to the function $y = e^x$, so we find that a (different) logarithm is the inverse function to $y = 10^x$. We say the inverse to 10^x is **logarithm to the base 10**, which we symbolize as $\log x$.

It is common to see $\log x$ written as $\log_{10} x$ to emphasize how the base differs from e. In fact the subscripted 10 here is superfluous since writing 'log' *means* a logarithm expressed in base 10. We will not employ the notation $\log_{10} x$ here.

When speaking aloud, we call log x either 'log ex' or 'log to the base 10 of ex'.

In the same way we can convert between e^x and ln x, so we can readily convert between $10x$ and $\log x$ since they are also inverse functions:

$$10^x = y \quad \text{so} \quad \log y = x \tag{11.2}$$

■ **Worked Example 11.7**

The pH of triflouroethanocic acid (**I**) is 4.5. From the definition of pH:

$$pH = -\log[H^+]$$

what is the concentration of the proton?

The Danish biochemist Søren Sørenson introduced this definition of pH in 1909.

We first rearrange the expression by multiplying both sides by -1:

$$-pH = \log[H^+]$$

We then remember how 10^x and $\log x$ are inverse functions. To remove the logarithm, we therefore perform the reverse operation, which is 10^x. We write:

$$10^{-pH} = 10^{(\log[H^+])}$$

But, once again, a function of its own inverse yields the original function. The right-hand side becomes merely $[H^+]$, so:

$$[H^+] = 10^{-pH}$$

Then, inserting numbers:

$$[H^+] = 10^{-4.5}$$

so:

$$[H^+] = 0.000\ 031\ 6\,\text{mol dm}^{-3}$$

This notation is quite difficult to read, so we generally prefer to write it in scientific notation (cf. Chapter 1) and say, $[H^+] = 3.16 \times 10^{-5}\,\text{mol dm}^{-3}$. ■

Some of the simpler values of log x can be calculated without a pocket calculate or computer, meaning we can convert from log x to 10^x without trouble. For example, we can say the log of 10^{-4} is -4; the log of 10^9 is 9, and so on. Alternatively, look at the values in Table 11.1.

Self-test 2

In each case, rearrange to make x the subject:

2.1	$\ln x = 7$	2.4	$e^x = 3$
2.2	$\ln x^2 = y$	2.5	$e^{(-6x)} = h$
2.3	$\ln xt = y$	2.6	$e^{(x^2)} = y$

The relationship between the logarithmic functions, ln and log

By the end of this section, you will know that:

- To convert from log x to ln x, multiply the value of log x by 2.303.

It is easy to obtain values of either log x or ln x using a pocket calculator or PC. However, it is sometime necessary when we perform algebra to convert between the two forms of logarithm.

In natural **base e**, the logarithm of 10 is 2.303. In other words, $e^{2.303} = 10$. This result has an important implication: to convert from a logarithm expressed in base 10 (log x) to a logarithm expressed in base e (i.e. a natural logarithm, ln x), we merely multiply the value of log x by 2.303.

■ **Worked Example 11.8**

Determine the values of both ln 45 and log 45. What is the value of their ratio?

Table 11.1 The relationship between factors of 10 and their logarithms expressed in base 10.

Number, x	log x	Rationale
1	0	because $10^0 = 1$
10	1	because $10^1 = 10$
100	2	because $10^2 = 100$
1000	3	because $10^3 = 1000$
$0.1 = 1/10$	-1	because $10^{-1} = 1/10^1 = 0.1$
$0.01 = 1/100$	-2	because $10^{-2} = 1/10^2 = 0.01$
$0.001 = 1/1000$	-3	because $10^{-3} = 1/10^3 = 0.001$

In fact, the ratio of *any* number expressed in both ln and $\log_{10}$ has a value of 2.303.

From an electronic calculator, we obtain the value of ln 45 as 3.806 66, and the value of log 45 as 1.605 32. Taking their ratio, we find:

$$\frac{\ln 45}{\log 45} = \frac{3.806\,66}{1.653\,21} = 2.303$$

The ratio of these two logarithms is 2.303. In fact, the ratio of *any* number expressed in both ln and log has the same value of 2.303. This generates an important result: to convert from log x to ln x, just multiply the value of log x by 2.303. ■

How logarithmic functions work

By the end of this section, you will know:

- The logarithm of a number greater than **1** is positive.
- A logarithm of **1** has a value of 0.
- The logarithm of a positive number less than **1** is negative.
- We cannot take the logarithm of a negative number.

We recall from Chapter 10 how, when multiplying together two algebraic power terms, we can predict the answer as a consequence of the mathematics. For example:

$$10^5 \times 10^4 = 10^{(5+4)}$$

In this section, we extend this idea to develop a powerful way of simplifying multiplication problems. We will use the logarithmic functions from the previous sections.

Imagine we wish to multiply together 8 and 5 to obtain 40. Obviously, we would usually just say, '$8 \times 5 = 40$'. We can look at the problem and write the answer straightaway. We cannot do so, however, if the numbers are more complicated.

The number 8 can be expressed as 10^x, the number 5 can be expressed as 10^y, and we can write 40 as 10^z. It should be clear how the values of x y, and z must each be different.

Following on from the laws of powers in the previous chapter, we can say:

$$10^x \times 10^y = 10^z \quad \text{and} \quad 10^{(x+y)} = 10^z$$

By writing the numbers this way, we are converting a multiplication problem into an addition problem. Therefore we have greatly simplified the problem.

We say that (in this example), x is the **logarithm** of 8, y is the logarithm of 5, and z is the logarithm of 40. Instead of $8 \times 5 = 40$, we could write:

$$
\begin{array}{ccccc}
8 & \times & 5 & = & 40 \\
\downarrow & & \downarrow & & \downarrow \\
10^x & \times & 10^y & = & 10^{x+y=z} \\
\downarrow & & \downarrow & & \downarrow \\
\log 8 & + & \log 5 & = & \log 40
\end{array}
$$

where the three log terms relate to the indices x, y, and x.

More usually, we employ the slightly abbreviated notation:

$$\log 8 + \log 5 = \log 40$$

We need to note:

- We could have written these logarithmic terms in any other base we choose.
- The expression is only correct if each term is expressed in the same base.
- If no subscript is indicated, we assume base 10.
- If ln rather than log, we are working with natural logarithms, expressed in base e.
- The correct citation style always uses lower-case letters: 'log'. We never write 'Log'.

Care: Most modern pocket calculators offer a variety of logarithmic functions. Be sure you press the right key!

■ **Worked Example 11.9**

Consider the simple esterification reaction:

for which equilibrium constant K has a value of 4 at 298 K. The value of K relates to the standard change in Gibbs function $\Delta G^{\ominus}$ according to the van't Hoff isotherm, $\Delta G^{\ominus} = -RT \ln K$ (see Worked example 11.4).

What is the value of $\ln K$ and what does this number actually relate to? ■

From a pocket calculator, the value of $\ln 4 = 1.386$.

This number means that e raised to the power of 1.386 (i.e. $e^{1.386}$) has a value of 4.

■ **Worked Example 11.10**

A solution of trifluoroethanocic acid (**I**) has a concentration of 10^{-4} mol dm^{-3}. What is the pH of the acid?

I

The definition of pH is:

$$pH = -\log_{10}[H^+]$$

where the 'H' derives merely from the symbol for hydrogen, so we always give it a big letter. The 'p' is the **compound operator**, $-1 \times \log_{10}$ of something. We always write it with a small letter. The phrase 'pH' means that we apply the operator 'p' to the domain, $[H^+]$. And the domain is abbreviated to just 'H'.

The p in 'pH' is short for *potenz*, the German word for 'power'.

To obtain the pH, we insert numbers into the equation:

$$pH = -\log_{10}[10^{-4}]$$

Clearly, the number 10 is raised to the power of -4 to obtain 10^{-4}, so the logarithm of 10^{-4} is -4. The pH of the aqueous solution of (**I**) is therefore:

It is usual to omit the positive sign at the end of this sum, and only cite a sign in the unusual cases when the pH is negative.

$$pH = -1 \times -4 = +4 \; ■$$

The laws of logarithms

By the end of this section, you will know:

- The sum of two logarithms ($\log a + \log b$) can be expressed as $\log (a \times b)$.
- The difference between two logarithms ($\log a - \log b$) can be expressed as $\log (a \div b)$.
- A logarithm of a power can be re-expressed such that $\log a^b = (b \times \log a)$.

While we have written these three laws in terms of log, they are equally valid for natural logarithms (ln), or indeed in terms of logarithms expressed in other bases.

Following directly from the way we defined logarithms above, we can formulate several laws.

$$\textbf{Law 1:} \quad \log a + \log b = \log(ab) \tag{11.3}$$

$$\textbf{Law 2:} \quad \log a - \log b = \log\left(\frac{a}{b}\right) \tag{11.4}$$

$$\textbf{Law 3:} \quad \log(a)^b = b \times \log a \tag{11.5}$$

■ **Worked Example 11.11**

Cis-2-butene (**II**) readily isomerizes to form *trans*-2-butene (**III**), reaction 2:

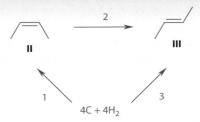

To determine the change in the Gibbs function for the formation of (**III**), we construct a simple Hess-law cycle and say $\Delta G_3^{\ominus}$ is the sum of the respective changes in the Gibbs function $\Delta G_1^{\ominus}$ and the $\Delta G_2^{\ominus}$:

$$\Delta G_3^{\ominus} = \Delta G_1^{\ominus} + \Delta G_2^{\ominus}$$

Use the van't Hoff isotherm to derive an expression for the equilibrium constant of reaction 3.

The van't Hoff isotherm says $\Delta G^{\ominus} = -RT \ln K$. So:

$$\Delta G_1^{\ominus} = -RT \ln K_1$$

$$\Delta G_2^{\ominus} = -RT \ln K_2$$

$$\Delta G_3^{\ominus} = -RT \ln K_3$$

Substituting for the $\Delta G^{\ominus}$ terms in the equations above, we obtain:

$$-RT \ln K_3 = -RT \ln K_2 + (-RT \ln K_1)$$

The factor of $-RT$ is common to each term, so we cancel throughout by $-RT$, leaving:

$$\ln K_3 = \ln K_2 + (\ln K_1)$$

Then, because we have the sum of two logarithms, we re-write the right-hand side using eqn (11.3):

$$\ln K_3 = \ln(K_2 \times K_1)$$

We can take the inverse of these logarithms, and discern the relationship:

$$K_3 = K_2 \times K_1 \quad \blacksquare$$

Self-test 4

Use eqn (11.3) to simplify each of the following:

4.1	$\ln 5 + \ln 3$	4.5	$\log gh + \log j$
4.2	$\ln 5 + \ln 8 + \ln 2$	4.6	$\log n + \log m + \log p$
4.3	$\log 20 + \log 7$	4.7	$\log_6 q + \log_6 r$
4.4	$\log_p a + \log_p 7b$	4.8	$\log_n 6 + \log_n 4t + \log_n 2f$

■ **Worked Example 11.12**

When deriving the Clausius–Clapeyron equation, the last-but-one step is:

$$\ln p_2 - \ln p_1 = -\frac{\Delta H^{\ominus}_{(vaporization)}}{R}\left(\frac{1}{T_2} - \frac{1}{T_1}\right)$$

where the two p terms are pressures and the two T terms are temperatures. R is the gas constant and $\Delta H^{\ominus}_{(vaporization)}$ is the mean enthalpy change on boiling. Simplify this equation using the laws of logarithms.

We first notice two logarithm terms, one of which is subtracted from the other. We therefore employ eqn (11.4). We rewrite the left-hand side of the equation as:

$$\ln\left(\frac{p_2}{p_1}\right) = -\frac{\Delta H^{\ominus}_{(vaporization)}}{R}\left(\frac{1}{T_2} - \frac{1}{T_1}\right)$$

which is the more usual form of the Clausius–Clapeyron equation. ■

> Notice how the first logarithm is placed on the *top* of the fraction (the numerator) and the second resides on the *bottom* (the denominator).

Self-test 5

Use eqn (11.4) to simplify each of the following:

5.1	$\ln 6 - \ln 3$	5.4	$\log y^2 - \log y$
5.2	$\ln 5f - \ln 5$	5.5	$\log 6g - \log 3g$
5.3	$\log 12 - \ln 4$	5.6	$\log h - \log p$

■ **Worked Example 11.13**

The so-called **quinhydrone electrode** relies on the redox reaction between quinine, Q (**IV**), and hydroquinone, H_2Q(**V**):

$$Q \quad O \qquad\qquad H_2Q \quad OH$$
$$\text{IV} \qquad\qquad\qquad \text{V}$$

Because the two species (**IV**) and (**V**) always remain in the solid state, the Nernst equation for this couple is simply:

$$E_{Q,H_2Q} = E^{\ominus}_{Q,H_2Q} + \frac{RT}{2F}\ln\left([H^+]^2\right)$$

where R is the gas constant, T is the thermodynamic temperature and F is the Faraday constant.

Simplify this expression using the laws of logarithms. ■

The concentration term within the logarithm is a square, so we are taking the logarithm of a power. Accordingly, we look at the logarithmic law in eqn (11.5). We can therefore re-write the logarithm term, saying:

$$E_{Q,H_2Q} = E^{\ominus}_{Q,H_2Q} + \frac{RT}{2F} \times 2 \times \ln([H^+])$$

We see how the factor against the proton concentration has moved, and is now positioned *outside* the logarithm, and written *before* it.

Now look more closely at the factors before the logarithm: the factor of 2 on the bottom of the fraction will cancel with the new factor of 2. Cancelling yields:

$$E_{Q,H_2Q} = E^{\ominus}_{Q,H_2Q} + \frac{RT}{\cancel{2}F} \times \cancel{2} \times \ln([H^+])$$

so:

$$E_{Q,H_2Q} = E^{\ominus}_{Q,H_2Q} + \frac{RT}{F}\ln([H^+])$$

■ **Worked Example 11.14**

Probably the most all-embracing definition of an equilibrium constant K is that defined in terms of a Pi product (see Chapter 2):

$$K = \Pi[\text{concentration}]^{v}$$

where v (the Greek letter nu) is the stoichiometric number, which we define as positive for products and negative for reactants.

Write the equilibrium constant K for the reaction:

$$aA + bB \rightarrow cC + dD$$

Then write an expression to describe the logarithm of K. ■

In this example, the small italic letters are the stoichiometric numbers, and the upright capitals are the participating chemical species.

We first write the equilibrium constant K:

$$K = [A]^{-a} \times [B]^{-b} \times [C]^{c} \times [D]^{d}$$

The two minus signs (before a and b) simply remind us that A and B are reactants. From Chapter 10, when we first looked at powers, we could have written K in its more familiar form as:

$$K = \frac{[C]^c [D]^d}{[A]^a [B]^b}$$

which illustrates the way that the bottom line of a fraction can be written as a number expressed to a negative power.

When we take the logarithm of this equilibrium constant K, it is generally easier to use two stages. Firstly we ignore the powers and leave them where they are. The first two laws of logarithms in eqns (11.3) and (11.4) help us obtain:

$$\ln K = \ln([A]^{-c}) + \ln([B]^{-b}) + \ln([C]^c) + \ln([D]^d)$$

Secondly, we use the third law of logarithms (eqn (11.5)). We say:

$$\ln K = -c \ln[C] - b \ln[B] + c \ln[C] + d \ln[D]$$

Self-test 6

Use eqn (11.5) to simplify each of the following:

6.1 $\log a + \log a + 3 \log a$

6.2 $\log b + \log b^2$

6.3 $\log c^3 + \log c$

6.4 $3 \ln c^2$

6.5 $6 \ln y + 2 \ln 4y$

Additional problems

11.1   The definition of pH is

$$pH = -\log[H^+]$$

Without using a calculator, what is the pH of a solution of HCl of concentration $10^{-5}\,\text{mol dm}^{-3}$?

11.2   The decay of a radioactive atomic nucleus follows an exponential relationship of the form:

$$\text{fraction remaining} = (1/2)^n$$

where n is the number of half-lives which have elapsed. Take the logarithm of both sides of the expression, and simplify it using the third law of logarithms.

11.3   Methyl ethanoate (**II**) is hydrolysed when dissolved in excess hydrochloric acid at 298 K. Its concentration [(**II**)] decreases with time during a chemical reaction, according to the first-order integrated rate equation:

$$\ln\left(\frac{[(\mathbf{II})]_0}{[(\mathbf{II})]_t}\right) = kt$$

where k is the rate constant and t is the time elapsing after the reaction has started. The initial concentration of (**II**) was $0.01\,\text{mol dm}^{-3}$, but 8.09×10^{-2} after 1260 seconds. Calculate the value of the rate constant, k.

11.4   The Arrhenius equation relates the rate constants of chemical reactions k and temperature T:

$$\ln\left(\frac{k_2}{k_1}\right) = -\frac{E_a}{R}\left(\frac{1}{T_2} - \frac{1}{T_1}\right)$$

The form of the **Arrhenius equation** in Worked example 11.2 is the so-called **linear form**; the version in this problem is the so-called **integrated form**.

where R is the gas constant and E_a is the activation energy. Using the laws of logarithms, expand the logarithm term on the left-hand side.

11.5   The **Debye–Hückel limiting law** relates the activity coefficient γ and a form of concentration known as ionic strength I:

$$\log \gamma = -A z^{+} z^{-} \sqrt{I}$$

where A, z^{+}, and z^{-} are constants. Rearrange the equation to make γ the subject.

11.6   The **Eyring equation** is a superior form of the Arrhenius equation, and relates the rate constants of chemical reactions k and temperature T:

$$\ln\left(\frac{k}{T}\right) = -\frac{\Delta H^{\#}}{RT} + \frac{\Delta S^{\#}}{R} + \ln\left(\frac{k_B}{h}\right)$$

where R is the gas constant, $\Delta H^{\#}$ is the enthalpy and $\Delta S^{\#}$ the entropy of activation, h is the Plank constant, and k_B is the Boltzmann constant. Using the laws of logarithms, expand the equation and then group together all the log terms.

11.7   The electrode potential $E_{Cu^{2+}, Cu}$ for the copper couple and the standard electrode potential $E_{Cu^{2+} \ominus, Cu \ominus}$ are related by the Nernst equation:

$$E_{Cu^{2+}, Cu} = E^{\ominus}_{Cu^{2+}, Cu} + \frac{RT}{2F}, \ \ln[Cu^{2+}]$$

where T is the temperature, R the gas constant, and F the Faraday constant. Rearrange the Nernst equation to make $[Cu^{2+}]$ the subject.

11.8   The **Clausius–Clapeyron equation** concerns gases and liquids at equilibrium, and relates the pressure p and temperature T:

$$\ln p_2 - \ln p_1 = -\frac{\Delta H^{\ominus}_{(vaporization)}}{R}\left(\frac{1}{T_2} - \frac{1}{T_1}\right)$$

The pressures are $p_1 = p^{\ominus}$ and $p_2 = 10 \times p^{\ominus}$. First use the laws of logarithms to rearrange the equation, then insert respective values of p to simplify further.

11.9   The **Tafel equation** relates the current I that flows when an electrode is immersed in a solution of analyte and the voltage is shifted by an amount η from its equilibrium potential:

$$\ln I = a + b\eta$$

where a and b are constants. Rearrange this equation to make I the subject.

11.10 Molecules sometimes leave solution to adhere to a solid surface. We say they 'adsorb'. If the adsorption interaction is weak, the amount leaving solution m depends on the concentration $[C]$ according to the **Freundlich adsorption isotherm**:

$$m = k[C]^{1/n}$$

where k and n are constants. Take the logarithm of this equation.

12

Powers III

Obtaining linear graphs from non-linear functions

Introducing the concept of linearization

By the end of this section, you will know:

- The rationale for linearizing curved graphs.
- The advantages of linearizing a curved graph.

It can be difficult to ascertain the equation of a curved line. One of the simplest and most powerful ways is to replot the data in such a way that a straight line is formed. We call this process **linearization**, because we make straight a graph that was previously curved.

Once the data are replotted in this way, we can then analyse the straight line and determine its formula in the normal way (see Chapter 8), i.e. determine its gradient and intercept. If we linearize a graph in a sensible way, the values of the gradient and intercept will tell us all we need to know in order to describe the original curved line.

To this end, we will see how to linearize graphs in which the exact nature of the function is already known, then explore ways of linearizing exponential function.

■ **Worked Example 12.1**

The so-called 'linear form' of the Gibbs–Helmholtz equation has the following form:

$$\frac{\Delta G^{\ominus}}{T} = \Delta H^{\ominus} \times \frac{1}{T} + c$$

where T is the thermodynamic temperature, $\Delta H^{\ominus}$ is the change in enthalpy, $\Delta G^{\ominus}$ is the standard change in the Gibbs function, and c is a constant. Obtain a linear graph from the data below, which relate to the reaction $NH_3 + \frac{5}{4}O_2 \rightarrow NO + \frac{3}{2}H_2O$.

T/K	250	300	350	400	450	500
$\Delta G^{\ominus}/kJ\,mol^{-1}$	−268	−240	−212	−184	−156	−129

A graph of $\Delta G^{\ominus}$ (as y) against T (as x) is not linear. This lack of linearity does not mean no relationship exists between $\Delta G^{\ominus}$ against T: the fact that the resultant curve is regular and smooth suggests there *is* a relationship. Rather, the curvature indicates the relationship is not simply $\Delta G^{\ominus} \propto T$.

> The symbol $\propto$ here means 'is proportional to'. It implies a *linear* proportion.

If we want to linearize these data, we must learn how to 'read' the Gibbs–Helmholtz equation cited in the question. The equation of a straight line is:

$$y = mx + c \tag{12.1}$$

If we compare the Gibbs–Helmholtz equation above with eqn (12.1), we see:

General case y $= m$ x $+c$

Specific case $\Delta G^{\ominus}/T$ $= \Delta H^{\ominus}$ $1/T$ $+c$

Therefore, the equation tells us that a graph of ($\Delta G^{\ominus}/T$) as y against $1/T$ as x should be linear. Figure 12.1 shows such a graph, which clearly is linear. ■

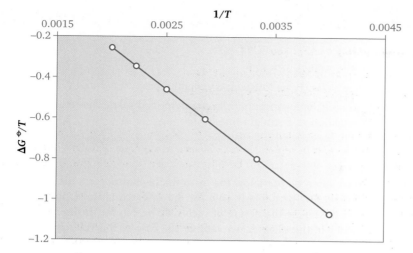

Figure 12.1 A linear graph drawn with data concerning $\Delta G^{\ominus}$ and T for the reaction $NH_3 + \frac{5}{4}O_2 \rightarrow NO + \frac{3}{2}H_2O$: a Gibbs–Helmholtz graph of $(\Delta G^{\ominus}/T)$ (as y) against $1/T$ (as x).

Self-test 1

Consider the following equations. In each case, how should we linearize them, i.e. how would we split them up to form an equation like $y = mx + c$?

1.1 The Nernst equation for the Cu^{2+},Cu couple:

$$E_{Cu^{2+},Cu} = E^{\ominus}_{Cu^{2+},Cu} + \frac{RT}{2F}\ln[Cu^{2+}]$$

where $RT/2F$ and $E^{\ominus}_{Cu^{2+},Cu}$ are both constants.

1.2 The linear form of the Clausius–Clapeyron equation:

$$\ln p = -\frac{\Delta H^{\ominus}_{(vap)}}{R} \times \frac{1}{T} + c$$

where $\Delta H_{(vap)}$, R, and c are constants.

1.3 The linear form of the second-order kinetic rate equation:

$$\frac{1}{[C]_{t=t}} = \frac{1}{[C]_{t=start}} + kt$$

where $[C]_{t=start}$ is the start concentration, and k is a constant.

1.4 Boyle's law relating the pressure and volume of a gas:

$$p = \frac{1}{V} \times \text{constant}$$

1.5 The linear form of the van't Hoff isochore:

$$\ln K = -\frac{\Delta H^{\ominus}_{(vap)}}{R} \times \frac{1}{T} + c$$

where $\Delta H^{\ominus}_{(vap)}$, R, and c are constants.

Linearizing curved graphs

By the end of this section, you will:

- Know that power functions follow two general types: $y = ba^x$ and $y = bx^a$.
- Know how to take the logarithm of the equation, and relate the reduced form to the equation of a straight line, $y = mx + c$.
- Know the best way to determine the values of the two constants a and b.

Unfortunately, we often do not know the exact mathematical form the graph should take. This observation is particularly true for the research chemist. In such a case, researchers must determine the nature of such relationship for themselves.

In the remainder of this chapter, we look at the case of functions having indices. In their most elementary forms, relationships involving indices follow one of two forms:

$$y = ax^b \qquad\qquad (12.2)$$
$$y = ab^x \qquad\qquad (12.3)$$

where x is the controlled variable, y is the observed variable, and a and b are constants.

In the laboratory, it is quite common to obtain data that clearly follow a relationship involving indices, yet we do not know whether they follow eqn (12.3) or eqn (12.2). Even if we know which equation the data follow, we do we know the values of a and b. The purpose of this chapter is to develop strategies which allow us to answer these questions.

Care: In a chemical context, the word **reduce** also means acquisition of an electron.

Our method will be to manipulate the data in such a way that the graphs become *linear*. We say we **linearize** the graph, or **reduce** it to linear form. We then relate this linear line with the equation of a straight line, $y = mx + c$. Generally, we draw a series of graphs until a choice of axes yields a linear graph. We then determine the values of the gradient m and intercept c, and relate them to the values of a and b.

Linearizing power equations of the type $y = ax^b$

By the end of this section, you will know:

- We first plot a graph of $\ln y$ (as y) against x.
- The gradient of the graph is b.
- The intercept is $\ln a$, so $a = e^{(\text{intercept})}$.

When seeking to linearize a graph of the type $y = ax^b$, we first take its logarithm:

$$\ln y = \ln a + \ln x^b$$

The last term can be rewritten using the third law of logarithms, and yields:

$$\ln y = \ln a + b \ln x \tag{12.4}$$

This equation has a clear resemblance to the equation of a straight line in eqn (12.1) above. Indeed, we can compare the two equations:

$$
\begin{array}{c|c|c|c}
\ln y = & \ln a + & b & \ln x \\
y = & c + & m & x
\end{array}
$$

We used natural logarithms **ln** in this example, but logarithms in base 10, $\log_{10}$, would have worked equally well.

In summary:

- We plot a graph of $\ln y$ (as the ordinate) against $\ln x$ (as the abscissa).
- We analyse this linear graph, saying the **gradient** $= b$.
- The intercept is $\ln a$, so the value of $a = e^{(\text{intercept})}$.

■ **Worked Example 12.2**

In an adsorption experiment, ethanoic acid **(I)** in water adsorbs weakly to the surface of alumina, Al_2O_3.

I

The mass of **(I)** adsorbed m is related to its concentration at equilibrium, $[C]$:

$[C]/10^{-4}\,\mathrm{mol\,dm^{-3}}$	6.8	13.9	22.7	37.5
$m/10^{-5}\,\mathrm{mol}$	6.08	8.97	10.8	13.7

These data follow the **Freundlich adsorption isotherm**:

$$m = k[C]^{1/n}$$

where m is the mass of substance adsorbed on the alumina, $[C]$ is the equilibrium concentration of substance adsorbed, and k and n are empirical constants.

By means of a suitable graph, determine the values of k and n. ■

We start by noting how the Freundlich equation has the form of eqn (12.2), i.e. $y = ax^b$. (In the Freundlich equation here, a is called k and b is called $1/n$.)

To obtain the values of a and b, we next take the logarithms of both m and $[C]$, and plot them. Figure 12.2 shows such a graph, with $\log x$ (as y) plotted against $\ln [C]$ (as x). While it shows a small amount of scatter, the graph is clearly linear. Its gradient is 0.47, and the intercept is -6.26.

When we compare the equation of a straight line (eqn (12.1)) with the logarithm of the exponential equation $y = ax^b$ (i.e. with eqn (12.4)), we see how:

- The value of the gradient $= 1/n$, so $1/n = 0.47$ and $n = 2.13$.
- The value of the intercept $= \ln k$, so $= \exp(\text{intercept})$, $k = e^{-6.26}$ and $k = 1.91 \times 10^{-3}$.

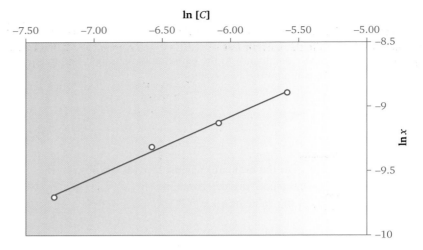

Figure 12.2 Freundlich isotherm plot of $\ln x$ (as y) and $\ln \log [C]$ (as x) constructed with data for ethanoic acid (**I**) adsorbed on alumina.

The equation of the line describing the mass of (**I**) adsorbing on alumina as a function of concentration is therefore:

$$m = 1.91 \times 10^{-3}[C]^{0.47}$$

Self-test 2

The viscosity η of a linear-chain polymer in solution increases with increases in the polymer's molar mass M, i.e. as the chain gets longer. The viscosity η and molar mass M are related according to the Mark–Houwink equation:

$$\eta = KM^{\alpha}$$

where K and α are constants which depend on the solvent and the polymer. The following data relate to polystyrene (**II**) dissolved in benzene:

$M/\text{g mol}^{-1}$	500	1000	2000	3000	3500	4000
$\eta/1000$	1.91	2.95	4.57	5.89	6.50	7.06

Using a suitable graph, determine values for K and α for polymer (**II**) in benzene.

Self-test 3

A rotated disc electrode (RDE) is shown in Fig. 12.3. The RDE rotates at a fixed frequency of ω. While the current I at the RDE depends on ω, the dependence is not linear.

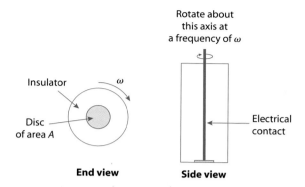

Rotate about this axis at a frequency of ω

Insulator

ω

Disc of area A

Electrical contact

End view   **Side view**

Figure 12.3 The RDE is spun about its central axis at a precisely known frequency ω, and the current I measured.

The data below were obtained at a RDE:

ω/Hz	20	40	60	80	100	120	140
I/mA	9.8	14.0	17.0	19.7	22.0	23.0	23.4

Using these data, plot a suitable graph to determine the correct relationship between I and ω (which is known as the **Levich equation**), i.e. what is n in the following equation?

$$I = k\omega^n$$

Linearizing exponential equations of the type $y = ab^x$

By the end of this section, you will know:

- To linearize such equations, we first plot a graph of $\ln y$ (as y) against x (and *not* $\ln x$).
- The gradient of the graph is $\ln b$, so a is $e^{(\text{gradient})}$.
- The intercept is $\ln a$, so $a = e^{(\text{intercept})}$.

When seeking to linearize a graph of the type $y = ab^x$, we first take its logarithm:

$$\ln y = \ln a + \ln b^x$$

The last term can be rewritten using the laws of logarithms to yield:

$$\ln y = \ln a + x \ln b \qquad (12.5)$$

Again, we remind ourselves of the equation of a straight line in eqn (12.1), $y = mx + c$. By comparing the two equations (12.1) and (12.5), we see:

$$\ln y = \begin{vmatrix} \ln a \\ \end{vmatrix} + \begin{vmatrix} x \\ x \end{vmatrix} \begin{vmatrix} \ln b \\ \end{vmatrix}$$
$$y = \begin{vmatrix} c \end{vmatrix} + \begin{vmatrix} x \end{vmatrix} \begin{vmatrix} m \end{vmatrix}$$

In summary:

- We plot a graph of $\ln y$ (as the ordinate) against x (as the abscissa), which will be linear.
- The gradient of this graph will be $\ln b$, so the value of $b = e^{(\text{gradient})}$.
- The intercept is $\ln a$, so the value of $a = e^{(\text{intercept})}$.

■ **Worked Example 12.3**

The pH of a solution of acid varies according to the concentration of solvated protons, $[H^+]$; see the table below:

$$[H^+] = 10^{-pH}$$

$[H^+]/\text{mol dm}^{-3}$	10^{-1}	10^{-2}	10^{-3}	10^{-4}	10^{-5}	10^{-6}	10^{-7}
pH	1	2	3	4	5	6	7

Use the data in the table to show this equation can be linearized by taking logarithms.

We first note how the equation follows the general type $y = ab^x$, where $a = 1$ and $b = 10$. Accordingly, a graph of $\ln [H^+]$ as y against pH (as x) should be linear. Figure 12.4 shows such a graph, which is clearly linear. Accordingly, we confirm the relationship has indeed been linearized. ■

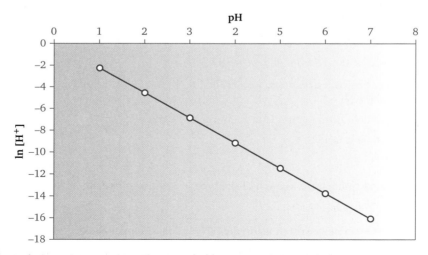

Figure 12.4 Graph of log $[H^+]$ as y against pH (as x).

Furthermore, we see how the gradient of this graph is 2.303 (see p.135). We can take this value as further confirmation, because $e^{2.303}$ has a value of 10.

■ **Worked Example 12.4**

A radioactive material decomposes. We define the half-life $t_{\frac{1}{2}}$ as the time necessary for exactly half of the material to decompose. If the original amount of material is n_0, the amount of material remaining after x half-lives is n_t, according to the equation:

$$n_t = n_0 \left(\frac{1}{2}\right)^{\left(\frac{t}{t_{1/2}}\right)}$$

where t is time. If we start with 1 mol of ^{60}Co, use the date below to determine the half-life $t_{\frac{1}{2}}$ for this radioisotope.

Time, t/min	10	20	30	40	50
n_t/mol	0.574	0.330	0.189	0.109	0.063

We first take the logarithm of this expression, to yield:

$$\ln n_t = \ln n_0 + \left(\frac{t}{t_{1/2}}\right) \times \ln\left(\tfrac{1}{2}\right)$$

We will rewrite this expression slightly:

$$\ln n_t = \ln n_0 + t \times \left(\frac{1}{t_{1/2}}\right) \times \ln\left(\tfrac{1}{2}\right)$$

so a graph of $\ln n_t$ (as y) against time t (as x) will be linear.

- The intercept will be $\ln n_0$.
- The gradient will be $\left(\frac{1}{t_{1/2}}\right) \times \ln\left(\tfrac{1}{2}\right)$. ■

Figure 12.5 shows such a plot, which is clearly linear.

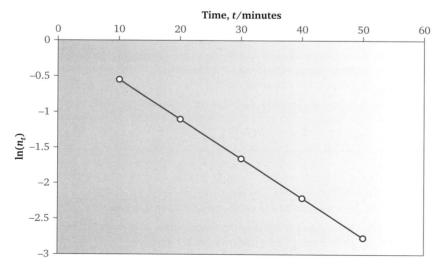

Figure 12.5 A radionuclide such as ^{60}Co is unstable, and decomposes spontaneously, so the amount of ^{60}Co$_t$ decreases with time, t: graph of $\ln n_t$ (as y) against time t (as x).

- The intercept is 0   The exponential of the intercept tells us the initial amount of material; we note how $\exp(0) = 1$, which confirms we started with 1 mol of ^{60}Co.
- The gradient is −0.0555   Manipulation of the equation above says the gradient equals: $\left(\frac{1}{t_{1/2}}\right) \times \ln\left(\frac{1}{2}\right)$

Therefore:

$$-0.0555 = \left(\frac{1}{t_{1/2}}\right) \times \ln\left(\frac{1}{2}\right)$$

Accordingly:

$$t_{1/2} = -\frac{1}{0.0555} \times -0.693$$

Therefore, the half life $t_{1/2}$ of ^{60}Co is 12.5 minutes.

Additional problems

12.1 In a kinetics experiment, a rate constant k varies with temperature T according to the linear form of the **Arrhenius equation**:

$$\ln k = -\frac{E_a}{R} \times \frac{1}{T} + c$$

where E_a and c are constants. How should data of k and T be linearized?

12.2 (Following on from question 12.1): consider the data below, which relate to the rate of removing a naturally occurring protein with bleach on a kitchen surface. What is the activation energy of reaction?

Temperature, $T/°C$	20	30	40	50	60
Rate constant, k/s^{-1}	2.20	2.89	3.72	4.72	5.91

Show these data can be linearized using the Arrhenius equation.

12.3 When an electrode is immersed in a solution of analyte and polarized, the time-dependent current I_t decreases with time t according to the **Cottrell equation**:

$$I_t = nFA[C]\sqrt{\frac{\pi D}{t}}$$

where n is the number of electrons, F is the Faraday constant, A is the area of the electrode, $[C]$ is the concentration of analyte, and D is the diffusion coefficient (a kind of velocity). How should data of I_t and t be linearized?

12.4 (Following on from question 12.3): the data in the following table relate to the one-electron reduction of an organic analyte at an electrode of area 0.21 cm^2, $[C] = 6.4 \times 10^{-7} \, mol \, cm^{-3}$, and with a diffusion coefficient of $4.0 \times 10^{-6} \, cm^2 \, s^{-1}$.

Time, t/s	1.0	2.0	3.0	4.0	5.0	6.0	7.0	8.0
Current, $I_t/\mu A$	46.0	32.5	26.5	23.0	20.6	18.8	17.4	16.3

Show these data can be linearized using the Cottrell equation.

12.5 During a first-order chemical reaction, the concentration of reactant $[A]_t$ decreases with time t according to the **first-order integrated rate equation**:

$$\ln[A]_t = kt + c$$

where k is the rate constant and c is a constant. How should data of $[A]_t$ and t be linearized?

12.6 (Following on from question 12.5): excess cerium(III) ion promotes the decomposition of hydrogen peroxide H_2O_2, according to a first-order rate law. The following data were obtained at 298 K.

Time, t/s	2	4	6	8	10	12	14	16	18	20
$[H_2O_2]_t/mol\ dm^{-3}$	6.23	4.84	3.76	3.20	2.60	2.16	1.85	1.49	1.27	1.01

Show the data can be linearized using the first-order integrated rate equation.

12.7 The electrode potential of the cadmium(II)–cadmium couple $E_{Cd^{2+},Cd}$ is related to the concentration of cadmium $[Cd^{2+}]$ according to the Nernst equation:

$$E_{Cd^{2+},Cd} = E^{\ominus}_{Cd^{2+},Cd} + \frac{RT}{2F} \times \ln[Cd^{2+}]$$

where $E^{\ominus}_{Cd^{2+},Cd}$ is the standard electrode potential, T is the temperature, R the gas constant, and F the Faraday constant. How should data of $[Cd^{2+}]$ and $E_{Cd^{2+},Cd}$ be linearized?

12.8 (Following on from question 12.7): linearize the following data set relating the concentration $[Cd^{2+}]$ and the electrode potential $E_{Cd^{2+},Cd}$.

$[Cu^{2+}]/mol\ dm^{-3}$	0.1	0.05	0.02	0.01	0.005	0.002
$E_{Cd^{2+},Cd}/V$	−0.430	−0.438	−0.450	−0.459	−0.468	−0.480

Show the data can be linearized using the Nernst equation.

12.9 The pressure p of gaseous iodine above solid iodine is a function of the temperature T according to the linear form of the Clausius–Clapeyron equation:

$$\ln p = -\frac{\Delta H^{\ominus}_{(vap)}}{R} \times \frac{1}{T} + c$$

where all other terms are constants. The following thermodynamic data refer to the sublimation of iodine.

$T_{(sublimation)}/K$	270	280	290	300	310	320	330	340
$p(I_2)/Pa$	50	133	334	787	1755	3722	7542	14 659

Linearize these data using the Clausius–Clapeyron equation.

12.10 It is rare for a reaction to follow third-order kinetics. The only known examples involve nitrogen dioxide, NO_2. In such cases, the concentration of NO_2 varies with time t according to the integrated third-order rate equation:

$$\frac{1}{[NO_2]_t^2} = kt + c$$

where k is the rate constant and c is a different constant. How should data of $[NO_2]$ and t be linearized?

13

Trigonometry

Introduction: naming a right-angled triangle

By the end of this section, you will:

- Learn the elementary nomenclature defining each part of a right-angled triangle.
- Know how to determine the sine, cosine, and tangent of a right-angled triangle.

Consider the right-angled triangle in Fig. 13.1. We need to learn a small amount of nomenclature. Firstly, we indicate the angles with capital letters and the sides with small letters.

The angles

- The internal angles are designated with capital letters.
- The sum of the three internal angles *A, B,* and *C* will always equal 180°.
- We write the Greek letter **theta** θ (pronounced *thay-tah*) if a single angle is indicated.
- Angles *A* and *B* can have any value between 0 and 90°. Since angle *C* is 90°, the other angles *A* and *B* will be θ, and $(90° - \theta)$.
- Angles less than 90° are said to be **acute**.
- By convention, the angle *A* is positioned opposite the side *a*, angle *B* is opposite the side *b*, and angle *C* is positioned opposite the side *c*.
- Some texts prefer to indicate angle *A* as *BAC*: to understand this nomenclature, we note that angle *A* lies between angles *B* and *C*, so angle B is therefore *ABC*; and angle *C* is *ACB*. We only really need this nomenclature when we go round a triangle in a clockwise or anti-clockwise direction, so we will not use it further.

The sides

- The angle of interest θ is properly called the **angle of focus**. We define it in terms of its position relative to the two sides that touch it.
- The side positioned next to the angle of focus is called the **adjacent**. We usually draw it *horizontally*.
- The side of the triangle that does not touch the angle of focus θ is generally drawn vertically. Its position explains why we call it the **opposite**.
- The longest side of the triangle is called the **hypotenuse**, and is generally drawn sloping at an angle, and touching θ at its lower end.

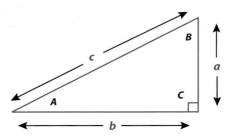

Figure 13.1 General depiction of a right-angled triangle.

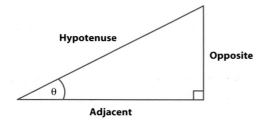

Figure 13.2 The names of the sides making up a right-angled triangle.

These names are shown schematically on Fig. 13.2.

The simple trigonometric functions

By the end of this section, you will:

- Learn the definitions of sine, cosine, and tangent.
- Know how to calculate a sine, cosine, and tangent.
- Know how to calculate an angle of focus θ if the sine, cosine, and tangent are already known.
- Learn the notation for trigonometric functions raised to powers.
- Learn to manipulate and rearrange algebraic phrases involving trigonometric functions.

Our definitions of sine, cosine, and tangent start with the three sides of a right-angled triangle.

- A **sine** is defined as the ratio of the adjacent to the hypotenuse:

$$\text{sine } \theta = \frac{\text{opposite}}{\text{hypotenuse}} \tag{13.1}$$

- A **cosine** is defined as the ratio of the adjacent to the hypotenuse:

$$\text{cosine } \theta = \frac{\text{adjacent}}{\text{hypotenuse}} \tag{13.2}$$

- A **tangent** is defined as the ratio of the adjacent to the hypotenuse:

$$\text{tangent } \theta = \frac{\text{opposite}}{\text{adjacent}} \tag{13.3}$$

Some people find it easier to remember the three definitions in eqns (13.1) – (13.3) using a simple rule: if sine = s, cosine = c, and tangent = t, and if opposite = o, adjacent = a and hypotenuse = h, then the three definitions above are summarized in the word:

From eqn (13.3), the gradient of the hypotenuse is its tangent.

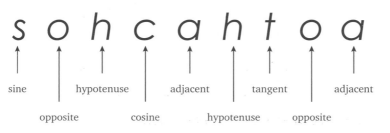

Generally, we don't write 'sine', 'cosine', or 'tangent', but abbreviate them respectively as 'sin', 'cos', and 'tan'.

■ **Worked Example 13.1**

A right-angled triangle has a hypotenuse of 5 cm, an adjacent of 4 cm, and an opposite of 3 cm. Calculate the values of sin θ, cos θ, and tan θ.

From eqn (13.1), the value of the sine θ is $\dfrac{3\,\text{cm}}{5\,\text{cm}} = 0.6$

From eqn (13.2), the value of the cosine θ is $\dfrac{4\,\text{cm}}{5\,\text{cm}} = 0.8$

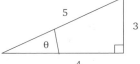

From eqn (13.3), the value of the tangent θ is $\dfrac{3\,\text{cm}}{4\,\text{cm}} = 0.75$ ■

■ **Worked Example 13.2**

The ions in a crystal lie in layered planes. The successive layers are separated by a distance d. We can determine the value of d with the technique of X-ray diffraction. Here, X-ray light of wavelength λ strikes the solid at an angle of θ, and is diffracted (see Fig. 13.3), according to the Bragg equation:

$$n\lambda = d\sin\theta$$

where n represents the number of diffractions.

Derive Bragg's law using the definition of sin θ in eqn (13.1).

We first look closely at Fig. 13.4, which represents a magnification of Fig. 13.3.

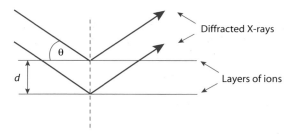

Figure 13.3 In the technique of X-ray diffraction, X-rays strike the sample at a glancing angle of θ degrees.

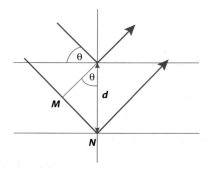

Figure 13.4 Inset of Fig. 13.3 highlighting the lengths involved during diffraction.

From the definition of a sine in eqn (13.1), $\sin \theta = $ (opposite $\div$ hypotenuse). The length of the hypotenuse is d and the length of the opposite is MN. Therefore,

$$\sin \theta = \frac{MN}{d}$$

Diffraction only occurs successfully when the length MN comprises an integral number of wavelengths, call it $n \times \lambda$. We therefore substitute for the length MN, saying it is $n\lambda$:

$$\sin \theta = \frac{n\lambda}{d}$$

Finally, we cross-multiply by d, to yield the Bragg equation, $d \sin \theta = n\lambda$. ∎

It is generally rare to calculate sin, cos, or tan using the respective values of adjacent, opposite, and hypotenuse. More usually, we know the angle of focus, and want to calculate a length from it.

If we know θ and the length of the hypotenuse (call it h), then we can calculate the length of both the adjacent and the opposite:

$$\text{length of the opposite} = h \sin \theta \qquad (13.4)$$

$$\text{length of the adjacent} = h \cos \theta \qquad (13.5)$$

■ **Worked Example 13.3**

A carbon–carbon single bond has a length of 154 pm. If the bond is positioned at an angle of 60° to a surface (see Fig. 13.5), what is the **projected length** of the bond?

To answer this question, we first note that the overall shape in Fig. 13.5 is that of a right-angled triangle. Accordingly, the projected length of the bond is the same as the adjacent of the triangle. We will therefore use eqn (13.5).

Projected length = adjacent = 154 pm × cos 60°

Projected length = 154 pm × 0.5

Projected length = 77 pm

So, if a bond is positioned obliquely to a surface at an angle of 60°, its apparent length is half its real length. ∎

A simple way of visualizing a **projected length** is to think, 'What would be the length of the shadow cast by the bond when illumined from a light positioned directly above it?'

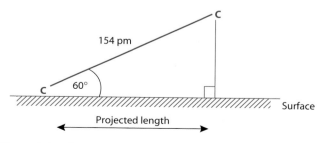

Figure 13.5 We can define the **projected length** as 'The perceived bond length when the bond is positioned at an angle to a surface or to another part of a molecule.'

....

Aside

To obtain the value of a sine, cosine, or tangent of an angle using a pocket calculator:

1. Ensure the calculator is set for *degrees*: at the top of the screen it will say 'DEG', 'RAD', or 'GRAD'. Press the DRG button until the display reads 'DEG'.

2. Enter the angle of focus as θ, which should normally be in degrees, °.

3. Press the appropriate function button, usually labelled sin , cos , or tan .

Steps 2 and 3 are reversed on some calculators.

....

■ **Worked Example 13.4**

A molecule of methyl viologen (**I**) adheres strongly to a platinum electrode. If the surface area of a flat molecule is 8800 pm², but the *projected* area is only 3800 pm², calculate the angle between the molecule of methyl viologen and the platinum.

$$H_3C - N^+ \bigcirc\!\!\!-\!\!\!\bigcirc N^+ - CH_3$$
$$2Cl^-$$

I

In this example, we need to perform a similar calculation to that in Worked example 13.3, but in reverse. We work again with eqn (13.5). The 'real area' corresponds to the hypotenuse and 'projected area' corresponds to the adjacent. Therefore:

$$\text{Projected area} = (\text{real area}) \times \cos\theta$$

Inserting numbers yields:

$$3800\,\text{pm}^2 = 8800\,\text{pm}^2 \times \cos\theta$$

We first rearrange by dividing throughout by 8800 pm²:

$$\frac{3800\,\text{pm}^2}{8800\,\text{pm}^2} = \cos\theta$$

so:

$$\cos\theta = 0.432$$

....

The **inverse function** to cosine θ is $\cos^{-1}\theta$. We occasionally write arccos θ. We *never* write 'inverse cos θ'.

....

To obtain the value of θ, we perform the inverse function to cosine, which we usually denote as $\cos^{-1}$. Therefore:

$$\theta = \cos^{-1}(0.432)$$

From a calculator or book of tables, the angle having a cosine of 0.432 is 64°. The molecule of methyl viologen (**I**) is inclined to the platinum surface at an angle of 64°. ■

■ **Worked Example 13.5**

It is usually difficult to obtain the NMR spectrum of *solid* samples. In order to analyse samples in the solid state rather than in solution, we use **magic-angle NMR**. Here, the tube containing the sample is spun fast while tilted at an angle of θ to the spectrometer's magnetic field (rather than perpendicular to it); the value of θ is given by the equation:

$$3\cos^2\theta - 1 = 0$$

Calculate the value of θ which satisfies this expression.

Before we solve this algebraic expression, we must introduce a new way of reading trigonometric functions. In this example, we see $\cos^2\theta$, which is an alternative way of writing $(\cos \theta)^2$; $\sin^3 \theta$ is the same as $(\sin \theta)^3$ etc. Most chemists agree that $\cos^2 \theta$ is not a logical way to write a square, but the usage is intended to help us distinguish between, say, $(\sin \theta)^2$ and $\sin \theta^2$. We will have to learn this notation style.

To solve this expression and make θ its subject, we first rearrange slightly:

$$3 \cos^2 \theta = 1$$

$$\cos^2 \theta = \frac{1}{3}$$

so:

$$\cos \theta = \sqrt{\frac{1}{3}} = 0.577$$

We then find the inverse cosine:

$$\theta = \cos^{-1}(0.577)$$

so:

$$\theta = 54.7°$$

The NMR sample should be spun at an angle of 54.7°. ∎

Self-test 2

Rearrange each of the following to obtain the value of θ:

2.1 $\sin \theta = 0.3$

2.2 $\cos \theta = 0.92$

2.3 $2 \cos \theta = 0.4$

2.4 $\frac{1}{2} \sin \theta = 0.45$

2.5 $\cos \theta + 2 = 2.1$

2.6 $\sin^2 \theta = 0.9$

2.7 $2 \tan^2 \theta = 9.0$

2.8 $4 \sin^3 \theta = 0.76$

Radians

By the end of this section, you will know:

- Degrees are not the only way of subdividing angles within a circle.
- There are 2π radians in a complete circle.

We are familiar with a circle comprising 360° (so a semi circle is 180°, a quadrant is 90°, and so on). Figure 13.6 shows these angles diagrammatically. In the alternative system of radians, we say a complete circle comprises 2π radians, where π is the familiar constant 3.142. One full circle therefore comprises 2×3.142 radians $= 6.284$ radians.

By the Greek letter π (pi), we mean the ratio of a circle's circumference to its diameter.

Help is at hand:

- Most calculators have the facility to work in radians rather than degrees. Look at your calculator's manual and see how to switch from degrees (probably labelled as 'DEG' on the LCD screen) to radians.

- Microsoft's program Excel automatically works in units of radians within trigonometric functions. For example, it will assume a function such as '= SIN(3)' means we want the sine of the angle, 'three radians', and will produce the result '0.14'.

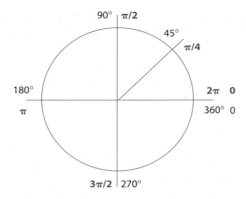

Figure 13.6 The conversion between radians and degrees. The radians are shown in bold print and the degrees in lighter print.

■ **Worked Example 13.6**

What is an angle of 22° in radians?

The easiest way to calculate radians from degrees is to employ ratios. We say, $360° = 2\pi$, so:

$$\frac{22°}{360°} = \frac{x}{2\pi}$$

and therefore:

$$x = \frac{22° \times 2 \times \pi}{360°}$$

that is:

$$x = \frac{44 \times 3.142}{360}$$

and:

$$x = 0.384 \text{ rad} \; ■$$

We have omitted the degree signs ° at this stage because the unit of 'degree' on the top and bottom will cancel.

Self-test 3

Convert each of the following from degrees to radians:

3.1 60°   **3.3** 160°

3.2 300°   **3.4** 34°

Pythagoras's theorem

By the end of this section, you will:

- Learn Pythagoras's theorem.
- Appreciate that Pythagoras's theorem requires a right-angled triangle.
- Learn that the theorem can be applied to two or three dimensions.
- Appreciate how Pythagoras's theorem is most useful to a chemist when looking at molecules and molecular arrangements in which a right-angled triangle is involved.

Pythagoras's theorem relates to right-angled triangles like that in Fig. 13.1. If we know the lengths a and b, we can calculate the length of the hypotenuse c according to eqn (13.6):

$$a^2 + b^2 = c^2 \qquad (13.6)$$

c here must be the longest side, but it does not matter which of a and b are the opposite and which the adjacent.

■ **Worked Example 13.7**

The original route to manufacturing nylon involved reacting adipic acid (**II**) with diaminohexame (**III**) to form a series of amide bonds. During the reaction, the carboxyl group of (**II**) comes level with the amine of (**III**) according to Fig. 13.7. In the transition-state structure, the distance between the carboxyl carbon and the amino nitrogen is 300 pm, and the length of the N–H bond is 103.8 pm long. Assuming the C–N–H bond is a right angle, what is the distance between the carboxyl carbon and the amino hydrogen (call it c)?

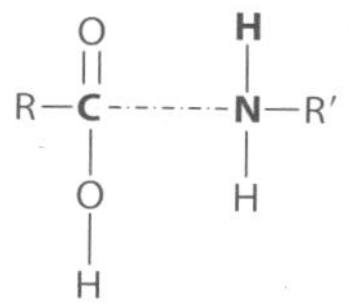

Figure 13.7 The carbon of the carboxyl must approach close enough to the nitrogen of the amine that electrons can transfer along the path of the dot-dashed line.

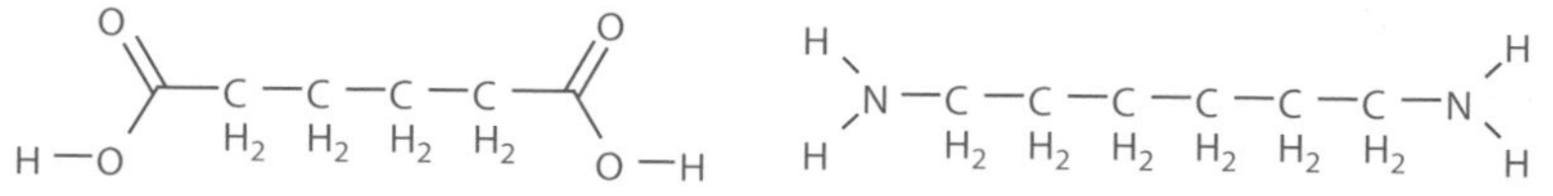

We will use Pythagoras's theorem in eqn (13.6), taking $a = 300$ pm and $b = 103.8$ pm.

$$c^2 = (300\,\text{pm})^2 + (103.8\,\text{pm})^2$$

so:

$$c^2 = 90\,000\,\text{pm}^2 + 10\,774\,\text{pm}^2$$

that is:

$$c^2 = 100\,774\,\text{pm}^2$$

Therefore:

$$c = \sqrt{100\,774\,\text{pm}^2} = 317\,\text{pm} \quad ■$$

Pythagoras's theorem can also be applied to a three-dimensional system.

■ **Worked Example 13.8**

Solid sodium chloride adopts a simple cubic structure when it crystallizes (see Fig. 13.8). If the shortest distance between adjacent chloride and sodium ions is 282 pm, then what is the next shortest distance, i.e. what is the length of the diagonal drawn on Fig. 13.6?

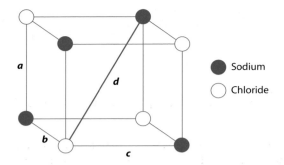

Figure 13.8 The ions in solid sodium chloride adoptsa simple cubic structure, with $a = b = c = 110\,pm$.

The angles between the sides a, b, and c are all right angles, so we employ an advanced form of the Pythagoras's theorem. The length of the diagonal d is given by eqn (13.7):

$$a^2 + b^2 + c^2 = d^2 \tag{13.7}$$

In this simple example, $a=b=c$, so we simplify slightly and say:

$$3a^2 = d^2$$

To make d the subject, we merely take the square root of both sides. We obtain:

$$d = \sqrt{3a^2}$$

Inserting numbers yields:

$$d = \sqrt{3 \times (282\,pm)^2}$$

so:

$$d = \sqrt{3 \times (79\,524\,pm^2)}$$

that is:

$$d = \sqrt{238\,572\,pm^2} = 488\,pm$$

We see how the diagonal is 73% longer than the sides a, b, and c. ∎

Self-test 4

Consider the following triangles using Pythagoras's theorem. In each case, assume the triangle has a right angle. The length c always relates to the *hypotenuse*.

4.1 $a = 2\,cm$ $b = 5\,cm$. What is c?   **4.2** $a = 7\,km$ $b = 13\,km$. What is c?

4.3   Consider the strained molecule, cyclobutadiene (**IV**): assuming all the internal angles are 90°, calculate the length of the *diagonal* in (**IV**). Take the bond lengths as C–C is 150 pm and C=C is 140 pm.

$$\begin{array}{ccc} HC & \!\!\!\!—\!\!\!\! & CH \\ \| & & \| \\ HC & \!\!\!\!—\!\!\!\! & CH \end{array}$$

IV

The cosine rule

By the end of this section, you will:

- Know the cosine rule.
- Know that the cosine rule does not require a right-angled triangle.

Sometimes we want to determine the properties of triangles that do not have a right angle. One of the more powerful equations to describe such triangles is the **cosine rule** (which is also called the 'cosine formula'). Again, it uses the letters from Fig. 13.1. It says:

$$c^2 = a^2 + b^2 - 2ab\cos C \tag{13.8}$$

When we use the cosine rule, we simply ensure that the angle of focus is C.

> Note how the cosine rule simplifies to yield Pythagoras's theorem if the angle of focus is 90°, because $\cos 90° = 0$.

■ **Worked Example 13.9**

Consider the amine $R\text{–}NH_2$ in Fig. 13.9. The angle between the two hydrogen atoms is 109.3°. The length of each N–H bond is 103.8 pm. What is the interatomic separation between the two hydrogen atoms?

The sides a and b here clearly have the same length. Side c of the triangle is the interatomic separation. Inserting values into eqn (13.8) yields:

$$\left(\begin{array}{c}\text{interatomic}\\\text{separation}\end{array}\right)^2 = (103.8\,\text{pm})^2 + (103.8\,\text{pm})^2$$

$$- 2 \times (103.8\,\text{pm}) \times (103.8\,\text{pm}) \times \cos 109.3°$$

$$\left(\begin{array}{c}\text{interatomic}\\\text{separation}\end{array}\right)^2 = 2 \times (103.8\,\text{pm})^2 - 2 \times \left((103.8\,\text{pm})^2 \times \cos 109.3°\right)$$

> The angle of focus in this triangle is greater than 90°. We say it is **obtuse**. An angle less than 90° is called **acute**.

This sum simplifies by factorising, to yield:

$$\left(\begin{array}{c}\text{interatomic}\\\text{separation}\end{array}\right)^2 = 2(103.8\,\text{pm})^2 \times [1 - \cos 109.3°]$$

so:

$$\left(\begin{array}{c}\text{interatomic}\\\text{separation}\end{array}\right)^2 = 21\,548\,\text{pm}^2 \times [1 - (-0.331)]$$

$$\left(\begin{array}{c}\text{interatomic}\\\text{separation}\end{array}\right)^2 = 21\,548\,\text{pm}^2 \times [1.331]$$

$$\left(\begin{array}{c}\text{interatomic}\\\text{separation}\end{array}\right)^2 = 28\,680\,\text{pm}^2$$

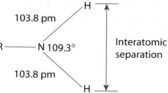

Figure 13.9 Interatomic separations are easily calculated with the **cosine rule**.

so:

$$\left(\begin{array}{c} \text{interatomic} \\ \text{separation} \end{array} \right)^2 = \sqrt{28\,680\,\text{pm}^2} = 169\,\text{pm} \quad \blacksquare$$

In fact, we could have avoided the cosine rule if we had bisected the obtuse triangle, to form two right-angled triangles, each with an angle of focus of $109.3° \div 2 = 54.65°$.

Self-test 5

Consider the following triangles. In each case, calculate the unknown variable using the cosine rule, eqn (13.8).

5.1

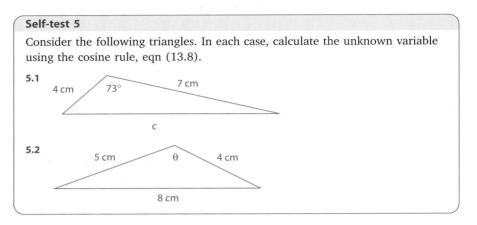

5.2

Additional problems

Questions 13.1–3 relate to the simplest of the hydrocarbons, methane (**V**), which is tetrahedral about the central carbon.

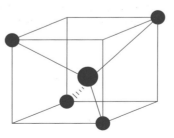

V

Conceptually, the simplest way to calculate the interbond angles in tetrahedral methane is to envisage the central carbon atom residing at the centre of a cube, with hydrogen atoms sitting on alternate vertices; see Fig. 13.10.

13.1 If the length of each side of the cube is a, calculate the length of a single C–H bond.

Figure 13.10 In a perfect **tetrahedral** arrangement, the central atom is located at the centre of a cube, and the four hydrogen atoms sit on alternate vertices.

13.2 If the length of each side of the cube is a, calculate the distance between the two hydrogen atoms at the top of the cube (i.e. determine the length of a diagonal across one of the cube faces).

13.3 Calculate the angle of focus θ, i.e. the bond angle $H-C-H$.

13.4 Consider the copper aquo ion (**VI**): assuming all angles are 90°, and the length of each $Cu-OH_2$ bond is the same at 167 pm, calculate the distance between two adjacent water molecules.

VI

13.5 Consider the cyclopentadienyl anion (**VII**), which is a regular pentagon: calculate the internal $C-C-C$ angle, expressing it both in degrees and in radians.

VII

13.6 Consider the transition-state structure (**VIII**) formed during the reaction of saponification of an ester with base:

VIII

The central carbon of (**VIII**) is nearly tetrahedral, with a $O-C-O$ bond angle of 112°. The length of the bond between the central carbon and the ionized oxygen is 139 pm, and the length of the bond between the central carbon and the methoxide is 125 pm. What is the distance between the two oxygen atoms?

13.7 Show how the cosine rule (eqn (13.8)) simplifies to form Pythagoras's theorem (eqn (13.6)) when the triangle is a right-angled triangle.

13.8 In a molecule of benzene (**IX**), what is the distance across the ring from carbon 1 to carbon 4? Assume the molecule is a perfect hexagon, each side having a length of 143 pm.

IX

13.9 (Following on from question 13.8): benzidine (**X**) is a powerful carcinogen. The bond between the two rings has a length of 153 pm. From Worked example 13.9, the angle between the two hydrogen atoms is 109.3° and the length of each N–H bond is 103.8 pm. The length of each N–H bond is 139.0 pm. What is the length of a molecule of benzidine (**X**)?

X

13.10 Consider the anti-cancer drug, *cis*-platin (**XI**), which is planar. Calculate the distance between the top Cl and NH_3 groups, assuming each angle is 90°, the Pt–Cl bond is 170 pm, and the Pt–N bond is 162 pm.

XI

14

Differentiation I

Rates of change, tangents, and differentiation

Introduction: tangents and the gradients of curves

By the end of this chapter, you will know:

- When we determine the gradient of a curve, we first draw a tangent.
- We determine the gradient of a tangent in the same way as for a straight line.

It is easy to determine the gradient of a *straight* line: we take the ratio of the distances travelled vertically (call it Δy) and horizontally (Δx). The gradient is then $\Delta y \div \Delta x$. The gradient is merely a rate of change, so the line changes a vertical distance Δy while simultaneously travelling a horizontal distance Δx.

Look at Fig. 14.1, which represents the amount of reactant remaining as a reaction proceeds. The curve has the mathematical form $y = e^{kt}$, where t is the time elapsing after the reaction commences, and k is a rate constant. The heavy straight line drawn against the curve represents the curve's gradient at a time of 1 minute: we call the line a **tangent**, and define it as 'a straight line that meets a curve at a point'.

A true tangent touches the curve only at one point—in this case at $x = 1$ minute—and never *crosses* the line. It's now easy to determine the gradient of the curve at $x = 1$: we determine the tangent's gradient by measuring Δx then Δy, and taking their ratio.

Self-test 1

Draw the following curves and then determine the gradient by first drawing a tangent:

1.1	$y = x^2$	Determine the gradient at $x = 3$.
1.2	$y = x^3$	Determine the gradient at $x = 4$.
1.3	$y = x^2 + 4$	Determine the gradient at $x = 2.5$.
1.4	$y = x^3 + x^2 + 2$	Determine the gradient at $x = 2$.

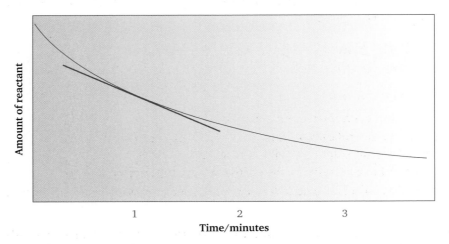

Figure 14.1 During a chemical reaction, the amount of reactant remaining (plotted as y) decreases as the reaction progresses, i.e. as a function of time (plotted as x): to determine the rate of reaction, we determine the gradient by drawing a tangent.

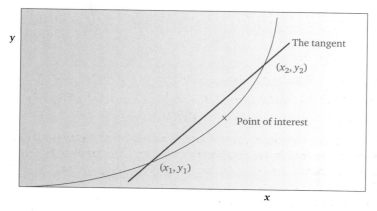

Figure 14.2 Determining the gradient of the tangent to the curve.

With a little practice, it becomes quite easy to determine the gradient of a curve. But problems remain with this approach. Firstly, we must note that a curve has an infinite number of gradients: the rate at which y changes will itself change as a function of x. In Fig 14.1, the gradient at $x=2$ is about half as steep as the gradient at $x=1$, the gradient at $x=3$ is about a quarter as steep, and so on. And there are also different gradients at $x=2.1$, 2.2, 2.3, etc., all of which differ.

The second problem is the way we actually measure the gradient of a tangent in practice. Most people find it quite hard to draw a tangent, so they are recommended to draw two points on the curve, placed at equal distances either side of the point where they want the gradient, like that in Fig. 14.2, then join the two points.

Having drawn a tangent according to Fig. 14.2, we determine a numerical value for the gradient saying:

$$\text{gradient} = \frac{\Delta y}{\Delta x} = \frac{y_2 - y_1}{x_2 - x_1} \qquad (14.1)$$

> This definition ensures the gradient has a *sign* as well as magnitude.

In practice, it does not matter if we write $(y_2 - y_1)$ or $(y_1 - y_2)$: what does matter is that the subscripted numbers agree, so if we write $(y_1 - y_2)$ then the bottom line of the fraction must be $(x_1 - x_2)$.

Introduction to differentiation: 'taking the limit'

By the end of this section, you will know:

- It is difficult to draw a tangent accurately.
- To counter this difficulty, the best tangent to a graph at a point X is obtained by taking two points, one either side of X, and joining them.
- A tangent becomes more accurate as the two points get closer together.
- When the two points coincide (one on top of the other), we say we 'take the limit'.

The method of determining a gradient outlined above will work well after a little practice, but it does have one major drawback. Unfortunately, the way we draw the heavy line can affect the value of the gradient. The numerical value of the gradient changes according to where we draw the tangent. Even if x_1 and x_2 are positioned at

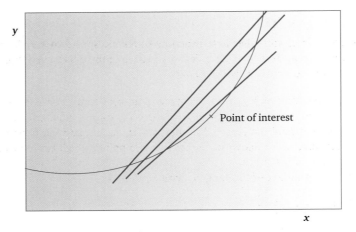

Figure 14.3 The magnitude of a tangent's gradient depends on where we draw it.

an equal distance either side of the point of interest, the actual distance chosen will itself change the value of the gradient. Look at Fig. 14.3.

Three possible gradients are drawn in Fig. 14.3. In each case, the tangent cuts the curve twice, with the two points of intersection being equidistant from the point of interest. The gradients of the three tangents are clearly very different. In fact the steepest (on the left) is about 40% steeper than the shallowest (on the right).

This situation is very unsatisfactory. We want to know the rate of change of the curve, and even drawing a tangent appears not to yield a good-quality gradient. After a moment's thought, we realize that the best gradient (by which we mean the most accurate) will be the one where the two points of intersection are closest to the point to interest. If the distance between the intersection points could be made infinitesimally small, then the gradient would be *exact*, and we would not need to worry any further.

....

Infinitesimal means '$1 \div \infty$', i.e. so small that its value is effectively zero.

....

Differentiation from first principles

By the end of this section, you will know:

- When differentiating, we employ algebra rather than drawing a graph.
- We start by taking two coincident points (x, y) and $(x + \delta x, y + \delta y)$, and write an expression for the gradient between them.
- The separation between the two points δx and δy is in fact zero.
- We give the name **differential coefficient** to the numerical value of the gradient.

As trainee mathematicians, we learn the gradient of the tangent becomes exact as *we take the limit*, i.e. as the separation between x_1 and x_2 becomes zero. When the separation between x_1 and x_2 is indeed zero, the tangent becomes a perfect tangent. And we can determine the gradient this way using algebra, which saves us the time and trouble of actually drawing a graph.

■ Worked Example 14.1

A molecule is accelerated inside the vacuum of a mass spectrometer. A graph of its velocity (as y) against time (as x) follows a curve of the type $y = x^2$. What is the gradient at the curve $y = x^2$ when $x = 4$?

We first draw the curve as Fig. 14.4. We draw a tangent that crosses the line at two points. We will call these two points (x, y) and $(x + \delta x, y + \delta y)$, where the Greek letter δ means an infinitesimal increment. In other words, we employ the following mathematical procedure, which allows us to consider two points which are in fact the same point, one on top of the other.

We obtain the gradient of the tangent with eqn (14.1):

$$\text{gradient} = \frac{\Delta y}{\Delta x} = \frac{y_2 - y_1}{x_2 - x_1}$$

Inserting the two points $(x + \delta x, y + \delta y)$ and (x, y) into this equation as (x_2, y_2) and (x_1, y_1) respectively, we obtain:

$$\text{gradient} = \frac{(y + \delta y) - y}{(x + \delta x) - x}$$

We then remember that the function of y is a square, $y = x^2$, so we can therefore substitute into the top line (the **numerator**) of the equation: y becomes x^2 and $(y + \delta y)$ becomes $(x + \delta x)^2$:

$$\text{gradient} = \frac{(x + \delta x)^2 - x^2}{(x + \delta x) - x}$$

We next multiply out the bracket on the top line of the equation, so:

$$\text{gradient} = \frac{(x^2 + 2x\,\delta x + \delta x^2) - x^2}{(x + \delta x) - x}$$

It is now time to simplify: on the bottom line, $(x + \delta x - x)$ becomes just δx, and on the top line the two x^2 terms also cancel. We are left with:

$$\text{gradient} = \frac{2x\,\delta x + \delta x^2}{\delta x}$$

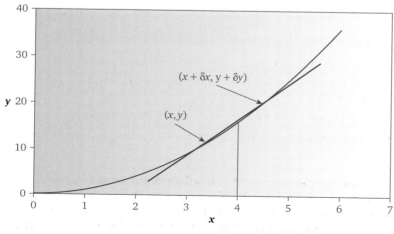

Figure 14.4 Determining the gradient to the curve $y = x^2$.

We can then simplify further, since there are δx terms on top and bottom. Perhaps it's easier if we first rewrite the fraction in two parts:

$$\text{gradient} = \frac{2x\,\delta x}{\delta x} + \frac{\delta x^2}{\delta x}$$

The two δx terms in the left-hand fraction will cancel, and the right-hand fraction becomes just δx. We are left with:

$$\text{gradient} = 2x + \delta x$$

And finally, we investigate how the value of the gradient simplifies further as the value of δx becomes infinitesimal, i.e. zero. In fact, the gradient of the curve $y = x^2$ becomes simply $2x$, because δx is 0.

By this procedure, we have **differentiated** the function $y = x^2$ to obtain a rather surprising result: whatever the value of x, the gradient of the curve is always $2x$. For example, the gradient when $x = 2$ is $2 \times 4 = 8$, the gradient at $x = 6$ is 12, and so on.

We give the name **differential coefficient** to the numerical value of the gradient obtained this way. ■

Self-test 2

Take the following functions and differentiate from first principles:

2.1 $y = x^2 + 3$ 　　　　　　　　2.3 $y = 4x^3$

2.2 $y = x^3$ 　　　　　　　　　　2.4 $y = x^3 + 4$

Important terminology

- We call the resultant gradient the **differential**, so the differential of $y = x^2$ is $2x$.
- We call the methodology employed in Worked example 14.1 **differentiation from first principles**.
- Instead of writing 'gradient' each time, we write $\frac{dy}{dx}$.
- The y on top and the x on the bottom will remind us of eqn (14.1). Instead of a capital Greek delta Δ on top and bottom, we write a small 'd', which stands for 'difference'. The difference 'd' implies an *infinitesimal* difference.

> ....
> Concerning this fraction, when reading aloud we would say, 'dee why by dee ex'.
> ....

Differentiation by rule

By the end of this section, you will:

- Know that it is possible to differentiate without the time-consuming necessity for determining an answer from first principles.
- Learn the simple rules for differentiation by rule.

Differentiating a function from first principles usually seems like a hard slog. We therefore want a simpler method of obtaining a gradient. Luckily we rarely ever need to differentiate from first principles. It's too time consuming and, being so tricky, we run the risk of making mistakes. From now on, we will differentiate—that is, obtain a gradient (a rate of change)—by a simpler route: we will differentiate by rule.

Look again at Worked example 14.1: we started with $y = x^2$ and obtained the differential as $2x$. Similarly, the differential of x^3 was $3x^2$. If we had differentiated x^4, we would have obtained a differential of $4x^3$. Hopefully we discern a pattern: when we differentiate a polynomial of the type x^n, we obtain $n \times x^{(n-1)}$. In mathematical notation we write:

$$y = x^n \qquad \frac{dy}{dx} = n \times x^{(n-1)} \qquad (14.2)$$

> This equation does not work when $x = 0$.

■ Worked Example 14.2

What is the differential of $y = x^5$?

Comparing this example with the template answer in eqn (14.2) shows the value of n is 5. Accordingly, the power will be $(5 - 1) = 4$. The differential is therefore $5 \times x^4$. In fact, we normally omit the multiplication sign and write the answer as just $5x^4$. ■

In many cases, we start with the x (or function of x) positioned within the *denominator*. In such cases, we first rewrite the function using the notation in chapter 10. For example, we would write $1/x^2$ as x^{-2}, $1/x^4$ as x^{-4}, etc. We then differentiate in the normal way via eqn (14.2).

■ Worked Example 14.3

According to Coulomb's law, the magnitude of the attractive force ϕ between two charges decays according to the inverse square of the intervening distance r. Mathematically, we say:

$$\phi = \frac{1}{r^2}$$

If we drew a graph of ϕ (as y) against r (as x), what would be the gradient, i.e. what is the differential of $\phi = 1/r^2$

> The version of **Coulomb's law** here is greatly simplified. In reality, the law includes several additional constant terms.

Methodology:

(i) Before we start, we recognize that we must rewrite this function without the fraction, i.e. in terms of scientific notation. We say $\phi = r^{-2}$.

(ii) To date, we have only employed the algebraic characters x and y. Because algebra is itself a form of shorthand, we are free to choose the shorthand notation of our choice. When we write a differential, we always write the variable of the vertical axis on the top as 'd variable' (here, as $d\phi$) and the variable of the horizontal axis on the bottom, so here we write dr. ■

Inserting values into the template in eqn (14.2), we say:

$$\phi = r^{-2} \quad \text{so} \quad \frac{d\phi}{dr} = -2 \times r^{(-2-1)}$$

The -2 at the beginning of the answer comes from the original power on r. The new power is -3 from $(-2 - 1)$. The answer is therefore:

$$\frac{d\phi}{dr} = -2\, r^{-3}$$

We could alternatively write this result as $-2/r^3$.

We need to note:

When we differentiate a function of the type x, we need first to write it as x^1. When we decrease the power by 1 (as before in accordance with eqn (14.2)), we obtain an answer of the form x^0. Now, since anything raised to the power of 0 has a value of 1, the differential of $y = x$ is actually 1. The gradient of the line x is 1.

Self-test 3

Differentiate the following by rule using eqn (14.2):

3.1 $y = x^6$ 3.6 $y = x^{-2}$

3.2 $y = x^{12}$ 3.7 $y = x^{-11}$

3.3 $y = x^5$ 3.8 $y = 1/x^7$

3.4 $y = x^3$ 3.9 $y = 1/x^3$

3.5 $y = x$ 3.10 $y = x^{2.73}$

■ Worked Example 14.4

In infrared spectroscopy, we watch bonds as they vibrate. The movement of vibrating atoms either end of a bond can be approximated to **simple harmonic motion (SHM)**, like two balls separated by a spring. The force f necessary to shift an atom or group away from the equilibrium positions follows the equation:

$$f = -kx^2$$

If we were to plot a graph of f (as y) against x (as x), what would its gradient be?

Methodology:

(i)   The word 'gradient' in the question alerts us that we want a rate of change.

(ii)   Because $f \propto x^2$, the graph will clearly be curved rather than linear, so we need to differentiate.

(iii) We cannot employ eqn (14.2) because we have a more complicated situation—the function of x has been multiplied by a constant (in this case, by $-k$). Accordingly, we employ a more realistic variant of eqn (14.2):

If $y = ax^n$, then

$$\frac{dy}{dx} = a \times n \times x^{(n-1)} \tag{14.3}$$

where a represents any constant.

Inserting values from the question into eq. (14.3), we obtain:

$$y = (-k)x^2$$

so:

$$\frac{dy}{dx} = (-k) \times 2 \times x^{(2-1)}$$

Omitting the multiplication signs, we write:

$$\frac{dy}{dx} = -2kx \quad ■$$

Self-test 4

Differentiate the following by rule, using eqn (14.3):

4.1 $y = 3x^3$

4.2 $y = 5x^{14}$

4.3 $y = 4x^5$

4.4 $y = 1.2x^{-7}$

4.5 $y = 4.3x^6$

4.6 $y = 10^6 x^{-4}$

■ Worked Example 14.5

The rotated-disc electrode (RDE) is one of the electrochemist's most useful analytical tools. We rotate such electrodes at a constant speed of ω and the resultant current is I. The Levich equation defines the relationship between I and ω:

$$I = k \times \omega^{1/2}$$

where k represents a jumble of constants. If we were to plot a graph of I (as y) against ω (as x), what would its gradient be?

> The value of k in the **Levich equation** accommodates many variables such as temperature T, number of electrons n, analyte concentration c, electrode area A, etc.

This example looks more complicated than those we saw earlier because the power is a fraction. In fact, we can still use eqn (14.3).

Methodology:

(i)   The word 'gradient' in the question alerts us that we want a rate of change.

(ii)   Because $I \propto \omega^{1/2}$, the graph will clearly be curved rather than linear, so we need to differentiate.

(iii) We remember that a square root can be expressed in terms of a decimal, so we write $\omega^{0.5}$.

Following point (iii), we write the abbreviated Levich equation as $I = k\omega^{0.5}$.
    To differentiate, we insert values into eqn (14.3), so:

$$\frac{dI}{d\omega} = 0.5k\omega^{-0.5}$$

The resultant power follows because $0.5 - 1 = 0.5$. ■

> In this example we have written I instead of y and ω instead of x, but will otherwise employ an identical methodology.

Whatever we do to the variables in an equation, constants do not change. That's why we call them constants. Consider the three curves $y = x^3$, $y = x^3 + 20$, and $y = x^3 + 40$, which are plotted in Fig. 14.5. The figure clearly shows the three lines in the graph are *parallel*.

Self-test 5

Differentiate the following by rule, applying eqn (14.3) in each case:

5.1 $y = 3x^3$

5.2 $y = 4x$

5.3 $y = 3.2x^{1/2}$

5.4 $y = 6 \times 10^5 x^2$

5.5 $y = 4.65x^{1/3}$

5.6 $y = 10x^{9.2}$

After a moment, we realize that parallel lines have the same gradients, but are merely shifted about the page. We see how the gradient at (say) $x = 2$ is the same on each graph.

A constant in an equation therefore merely shifts the line up or down a graph, and in no way alters the gradient. *For this reason, we obtain a zero if we differentiate a constant.*

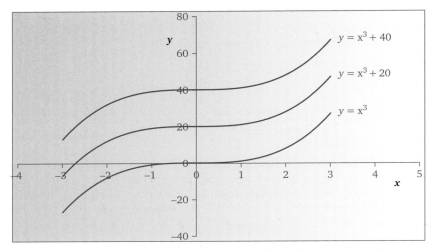

Figure 14.5 Graph showing the three parallel curves $y = x^3$, $y = x^3 + 20$, and $y = x^3 + 40$.

■ **Worked Example 14.6**

Λ^0 is the conductivity at a single concentration and can have only one value. Accordingly, Λ^0 is a constant.

The conductivity Λ of a simple ionic salt in solution follows the so-called Onsager equation:

$$\Lambda = \Lambda^0 - b\sqrt{[C]}$$

where b is a constant, $[C]$ is the concentration, and Λ^0 is the limiting conductivity, i.e. it is also a constant. If we were to plot a graph of Λ (as y) against $[C]$ (as x), what would its gradient be?

Methodology:

(i)   The word 'gradient' in the question alerts us that we want a rate of change.

(ii)   Because $\Lambda \propto [C]^{\frac{1}{2}}$, the graph will clearly be curved, so we need to differentiate.

(iii) We will differentiate each term, one at a time.

(iv)   We will treat the Λ^0 term as a constant, so its differential is zero.

We first rewrite the equation slightly, as $\Lambda = \Lambda^0 + (-b)[C]^{0.5}$. The Λ^0 term does not vary because it is a constant, so it vanishes when we differentiate. The term $(-b)[C]^{0.5}$ is a simple function, and can be approached in the same way as other examples:

$$\frac{d\Lambda}{dc} = 0 + 0.5 \times (-b)[C]^{-0.5}$$

so:

$$\frac{d\Lambda}{dc} = -0.5b[C]^{-0.5}$$

We could have written this answer in a variety of other ways:

$$\frac{d\Lambda}{d[C]} = -\frac{0.5\,b}{[C]^{0.5}} \quad \text{or} \quad \frac{b}{2[C]^{0.5}} \quad \text{or} \quad \frac{b}{2\sqrt{[C]}}$$

Each of these three alternatives is equally valid. ■

Self-test 6

Differentiate the following by rule:

6.1 $y = x^3 + 12$

6.2 $y = 2x^2 - 3$

6.3 $y = 6 - x^{\frac{1}{2}}$

6.4 $y = 8 + \dfrac{1}{x^3}$

6.5 $2y = x^2 - 2$ (remember first to 'reduce the equation' to yield an equation of the form $y = \ldots$)

■ **Worked Example 14.7**

The potential of a saturated calomel electrode (SCE) varies with temperature. We can calculate its potential E_{SCE} with a power series:

$$E_{SCE} = 0.242 - (7 \times 10^{-3})T + (4 \times 10^{-7})T^2$$

What is the gradient of a graph of E_{SCE} (as y) against temperature T (as x)?

Methodology:

(i) The word 'gradient' in the question alerts us that we want a rate of change.

(ii) Because E_{SCE} is not a linear function of T, the graph will clearly be curved rather than linear, so we need to differentiate.

(iii) This example looks more complicated than those previously because the right-hand side has three terms rather than just one. We differentiate the expression by differentiating one term at a time.

$$E_{SCE} = 0.242 - (7 \times 10^{-3})T + (4 \times 10^{-7})T^2$$

$$\downarrow \qquad\qquad \downarrow \qquad\qquad\qquad \downarrow$$

$$\frac{dE_{SCE}}{dT} = 0 \quad - (7 \times 10^{-3}) \quad + (4 \times 10^{-7}) \times 2T$$

so:

$$\frac{dE_{SCE}}{dT} = -(7 \times 10^{-3}) + (8 \times 10^{-7})T \ ■$$

Self-test 7

Differentiate the following by rule, eqn (14.3):

7.1 $y = x^4 + x^2$

7.2 $y = x^3 - x^{-4}$

7.3 $y = x^7 - 6x^2 + 2$

7.4 $y = \sqrt{x} + 12x - 3$

Additional problems

14.1 A molecule is moving through the high vacuum of a mass spectrometer at a velocity of v, defined as:

$$v = \frac{l}{t}$$

where l is the distance travelled in a time t. The particle is accelerated electromagnetically. Write an expression for the acceleration a, dv/dt.

14.2 In the electrochemist's technique of cyclic voltammetry, the peak current I is a simple function of the scan rate v, according to the **Randles–Sevčik equation**:

$$I = 0.4463nF \left(\frac{nF}{RT} \right)^{1/2} D^{1/2} [C] v^{1/2}$$

where n is the number of electrons transferred, F is the Faraday constant, D the diffusion coefficient, and $[C]$ the concentration of analyte. Write an expression to describe the gradient of a graph of I (as y) against v (as x), i.e. dI/dv. [Hint: treat the equation as $I = kv^{1/2}$]

14.3 The pressures at the solid–liquid **phase boundary** for propane are given by the empirical expression:

$$p = -718 + 2.386 \times T^{1.283}$$

Write an expression for the gradient of this phase boundary, i.e. of dp/dT.

Empirical means 'obtained from experimental observation'.

14.4 Molecules sometimes leave solution to adhere to a solid surface. We say they 'adsorb'. When the adsorption is weak, the amount adsorbing m depends on the concentration $[C]$ according to the **Freundlich adsorption isotherm**:

$$m = k[C]^{1/n}$$

where k and n are constants. Write an expression for $dm/d[C]$.

14.5 The speed at which electrons move through a semiconductor relates to its mobility μ. The temperature dependence of μ is complicated, but for many semiconductors, μ shows the following relationship:

$$\mu = k\,T^{3/2}$$

where k represents a collection of constants. Write an expression for $d\mu/dT$.

14.6 In an atom, the electrons circulate around a central nucleus. Because the atom contains several electrons, an electron far from the nucleus can be screened and experience a lesser interaction with the nucleus. The **effective potential of the nucleus** Z_{eff} of the nucleus is given by the expression:

$$Z_{\text{eff}} = -\frac{Ze^2}{4\pi\varepsilon_0 r} + \frac{l(l+1)\hbar^2}{2\mu r^2}$$

where r is the distance of the electron from the nucleus, and all other terms are constants. Write an expression for dZ_{eff}/dr.

14.7 Light is scattered as it passes through a solution containing microscopically small particles. For example, milk looks white because it contains a suspension of tiny spheres of insoluble oils and fats. The intensity of the scattered light I relates to the wavelength of the light scattered λ:

$$I = I_0 \frac{\pi \alpha^2}{\varepsilon_r^2 \lambda^4 r^2} \sin^2 \phi$$

where all the other terms may be considered to be constant. Write an expression for $dI/d\lambda$.

14.8   The ideal-gas equation ($pV = nRT$) often fails to describe the behaviour of real gases. One of the better alternatives is the so-called virial equation:

$$pV_m = RT\left[1 + \frac{B}{V_m} + \frac{C}{V_m^2}\right]$$

where V_m is merely $V \div n$. Rearrange the equation slightly, then write an expression for dp/dV_m.

14.9   If two electric dipoles are aligned in a parallel arrangement, the potential energy V of their interaction is given by the expression:

$$V = \frac{\mu_1 \mu_2}{2\pi\varepsilon_0 r^3}$$

where r is the inter-dipole separation, and all the other terms are constants. Write an expression for dV/dr.

14.10   Worked example 14.6 described the Onsager equation:

$$\Lambda = \Lambda^0 - b\sqrt{[C]}$$

This equation is empirical, but it can be derived from first principles using the Debye–Hückel laws. In this way, the constant b has the form:

$$b = \frac{qz^3\varepsilon F}{24\pi\varepsilon_0 RT}\left[\frac{2}{\varepsilon RT}\right]^{1/2}$$

where z is the charge on the ions, T is the temperature, and other terms are constants. Combine the T terms together, and then write an expression describing the temperature dependence of b, i.e. write an expression for db/dT.

15

Differentiation II

Differentiating other functions

Differentiating more functions

By the end of this chapter, you will know:

- How to differentiate functions such as exponentials, logarithms, sines, and cosines.
- How to 'read' these functions; that is, see what the operator, what the domain, and what the factor(s) are.

There are a large number of functions we cannot differentiate 'by rule', using the formula in the previous chapter. The simplest are exponential functions of the form e^{ax}.

Exponential functions

When we differentiate an exponential of the type $y = e^{ax}$ (where a is a constant) we employ the relationship in eqn (15.1):

$$\frac{dy}{dx} = a\,e^{ax} \qquad\qquad (15.1)$$

Nomenclature:

- The exponential is an **operator**.
- The operator operates on a **domain**: the domain represents the set of possible values of an independent variable. In the example above, the domain is ax.
- The final differential has been multiplied by a **factor** of a.

We need to note:

- The domain often needs to be written within brackets.
- The differential of the exponential *always* contains the original exponential, whatever its domain. Neither the operator nor the domain changes: in fact, only the factor that precedes the exponential will change.

■ **Worked Example 15.1**

What is the gradient of the exponential function $y = e^{4x}$?

Methodology:

(i) The word 'gradient' in the question alerts us that we want a rate of change.

(ii) We insert respective terms into eqn (15.1). As $y = e^{4x}$, so:

$$\frac{dy}{dx} = 4\,e^{4x} \quad ■$$

■ Worked Example 15.2

1,2-dichloroethane (**I**) decomposes in the gas phase at 780 °C with a rate constant k of $4.4 \times 10^{-3} s^{-1}$.

The amount of (**I**) at any time during the course of the reaction is $[(\mathbf{I})]_t$, and equals:

$$[(\mathbf{I})]_t = [(\mathbf{I})]_0 \, e^{-kt}$$

where t is the time since the reaction commenced. The term $[(\mathbf{I})]_0$ represents the concentration of reactant at the start of the reaction, so it represents a constant. What is the rate of reaction as a function of time?

By asking for the 'rate of reaction', we are really asking for the differential. For simplicity, we rewrite the equation slightly, $c_t = c_0 \, e^{-kt}$, to make it look less intimidating:

$$\frac{dc_t}{dt} = (-k) \times c_0 \, e^{-kt}$$

We then reinsert the concentration c_0 and the numerical value of the rate constant k, saying:

$$\frac{dc_t}{dt} = (-4.4 \times 10^{-3}) \times [(\mathbf{I})]_0 \, e^{(-4.4 \times 10^{-3})t}$$

This example is marginally more complicated than Worked example 15.1 because:

- The function was itself multiplied by a constant (in this case, a concentration).
- It illustrates the way we often simplify an equation by first substituting.   ■

In general, if $y = be^{ax}$, then:

$$\frac{dy}{dx} = (a \times b) \, e^{ax} \qquad (15.2)$$

Self-test 1

Differentiate each of the following exponential functions using eqn (15.2):

1.1 $y = e^{5x}$

1.2 $y = e^{3.4x}$

1.3 $y = 7e^{fx}$

1.4 $y = 9.3e^{4.2x}$

1.5 $y = d \, e^{dx}$

1.6 $y = 8.73 \, e^{(4 \times 10^{-7})x}$

1.7 $y = e^x - e^{-x}$

1.8 $y = 3x^4 - 7x^2 + e^{5x}$

Logarithmic functions

The differential of a logarithmic function is not intuitively obvious. If $y = \ln ax$, then:

$$\frac{dy}{dx} = \frac{1}{x} \qquad (15.3)$$

We need to notice how the constant of a disappears completely.

■ **Worked Example 15.3**

What is the differential of $y = \ln 7x$?

Inserting values into the template in eqn (15.3):

$$\frac{dy}{dx} = \frac{1}{x}$$

Notice that the factor of 7 has disappeared because the function in the question was a logarithm. ■

Rationale

The rationale for the disappearing constant in eqn (15.3) comes from the laws of logarithms, which says we can split the expression $\ln 7x$ into two. We say:

$\ln 7x = \ln 7 + \ln x$

Because 7 is merely a number, $\ln 7$ is also just a number—in other words, $\ln 7$ is a constant, and the differential of a constant is always zero.

■ **Worked Example 15.4**

The relationship between pH and hydrogen ion concentration is given by the equation:

$$pH = -\log_{10}[H^+]$$

If we made a series of solutions of the strong acid trifluoroethanoic acid (**II**), and measured the gradient of a graph of pH (as y) against concentration of the proton $[H^+]$ (as x), what would its gradient be?

II

Methodology:

(i) Before we can differentiate this example, we must convert from a logarithm in base 10 (i.e. $\log_{10}$) to a natural logarithm (i.e. ln). From page 135, we say $\log10$ $x = 2.303 \ln x$, so:

$$pH = -2.303 \times \ln[H^+]$$

The differential in eqn (15.4) is effectively $b \times (1/x)$.

(ii) We then differentiate as normal, using an extended version of eqn (15.4). If $y = b \times \ln ax$, then:

$$\frac{dy}{dx} = \frac{b}{x} \qquad (15.4)$$

Inserting terms into the template differential in eqn (15.4), we obtain:

$$\frac{dpH}{d[H^+]} = \frac{2.303}{[H^+]} \quad ■$$

■ **Worked Example 15.5**

In 1905, the electrochemist Julius Tafel devised one of the simplest models to describe the rate of an electron transferring between an analyte and an electrode. The equation he derived still bears his name:

$$I = a + b \ln \eta$$

The Greek letter η is called 'eta'. We pronounce it as 'aye-tah'.

where I is the current measured as a consequence of the electrons transferring, and η is the deviation of the potential away from its equilibrium value, which we call the overpotential. a and b are simply constants.

What is the rate of change of a graph of I (as y) against η (as x)?

Methodology:

(i)   We will differentiate the Tafel equation one term at a time in just the same way as we differentiated an equation incorporating a number of simple polynomial terms.

(ii)   The second part of the Tafel equation is a logarithm, so we insert terms into eqn (15.4):

$$\frac{\mathrm{d}I}{\mathrm{d}\eta} = 0 + b \times \frac{1}{\eta}$$

which we would normally write as:

$$\frac{\mathrm{d}I}{\mathrm{d}\eta} = \frac{b}{\eta} \quad \blacksquare$$

Self-test 2

Differentiate each of the following logarithmic functions using eqns (15.3) and (15.4):

2.1	$y = \ln x$	**2.6**	$y = ab \ln gx$
2.2	$y = \ln 7x$	**2.7**	$y = 4x - 3 \ln x$
2.3	$y = 5.4 \ln x$	**2.8**	$y = x^4 - \dfrac{1}{x^3} + \ln 4x$
2.4	$y = 45 - \ln x$	**2.9**	$y = 1000x + \ln 3x$
2.5	$y = \ln(4 \times 10^{-6})x$	**2.10**	$y = \ln ax - c \ln bx$

We need to note:

- We again call the function written to the right of the symbol 'ln' its 'domain'. The domain often needs to be written within brackets.

- The domain of the logarithm often remains unaltered following differentiation.

Differential coefficients of trigonometric functions

By the end of this section, you will:

- Recall that angles may be expressed in radians rather than degrees.

- Know that we cannot (readily) obtain a differential coefficient of a sine or cosine unless we work in **radians**.

The general name **trigonometry** (see Chapter 13) describes the mathematics of sines cosines, etc. Before we can learn the correct way to differentiate trigonometric functions, we need to consider a technicality. Unfortunately, the differential of a sine or cosine is far from straightforward: each differential of a trigonometric function would include a complicated additional term. To avoid such terms—and to save us the additional work of learning how to derive them—we will not work in terms of degrees, but **radians**.

There are 2π radians in a circle, so an angle of 360° is equivalent to 2π radians, 180° is equivalent to π radians, 90° equates to $\pi/2$, and so on.

Differentiating trigonometric functions

We can now consider the differentiation of trigonometric functions.

Figure 15.1(a) shows the shape of the simple function $y = \sin x$, and Fig. 15.1(b) shows the differential of $y = \sin x$. In being a rate of change, the shape of Fig. 15.1(b) is in fact the differential of $y = \sin x$.

Figure 15.2(a) shows the simple function $y = \cos x$, and Fig. 15.2(b) shows the differential of $y = \cos x$. We now look critically at the four graphs on pages 188 and 189. The differential of $y = \sin x$ looks identical to $\cos x$ (i.e. Figs 15.1(b) and 15.2(a), respectively, look the same). In fact, provided we work in radians, the two graphs *are* identical. Again, the differential of $y = \cos x$ looks almost identical to $y = \sin x$: they are indeed almost identical if we work in radians rather than degrees. The only difference between the two graphs is their sign, because the shape of the differential in Fig. 15.2(b) is the *mirror image* in the horizontal direction of the function $y = \cos x$ in Fig. 15.1(a).

We can express these relationships mathematically by saying the differential of a sine generates a function of a cosine, and vice versa:

$$y = \sin ax \qquad \frac{dy}{dx} = a \cos ax \qquad\qquad (15.5)$$

$$y = \cos ax \qquad \frac{dy}{dx} = -a \sin ax \qquad\qquad (15.6)$$

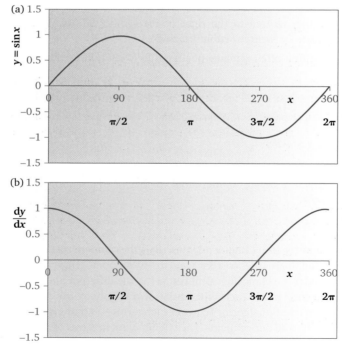

Figure 15.1 (a) The function $y = \sin x$, and (b) its differential with respect to x. Degrees are included for reference only: the value of x must be expressed in radians.

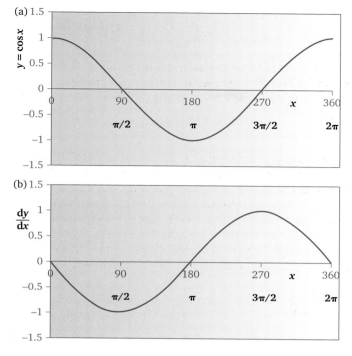

Figure 15.2 (a) The function $y = \cos x$, and (b) its differential with respect to x. Degrees are

We need to note:

- Again, the sine or cosine operator each operates on a domain. The domain often needs to be written within brackets.
- We always write the domain to the right of the operators 'sin' or 'cos'.
- The domains of the sine or cosine functions *always* remain unaltered. This is a universal rule: we *always* retain the domain, whatever its complexity.
- The differentiation of either a sine or of a cosine will follow the same rules in just about every way. The only real exception is the way the differential of a sine is a cosine, but the differential of a cosine is a sine multiplied by -1.

■ Worked Example 15.6

In X-ray crystallography, the Bragg equation allows us a simple means of determining the distance d between successive layers in a crystal (see Fig. 15.3). In its simplest form, it relates the X-rays' wavelength λ, the number of reflections n, and the angle through which the X-rays are scattered, θ, as:

$$\lambda = \frac{2d}{n} \sin \theta$$

What is the rate of change of λ with θ?

Inserting terms into the template in eqn (15.5), we obtain:

$$\frac{d\lambda}{d\theta} = \frac{2d}{n} \cos \theta$$

We obtained this result via eqn (15.5), which implies that we are using θ in radians. ■

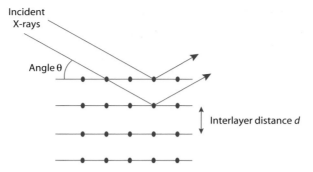

Figure 15.3 A beam of X-rays diffracts when striking a crystal with a repeat lattice.

■ **Worked Example 15.7**

What is the differential of the equation $y = 4\cos(-5x)$?
Inserting terms into the template in eqn (15.4), we obtain:

$$\frac{dy}{dx} = -4 \times (-5)\sin(-5x)$$

so:

$$\frac{dy}{dx} = +20\sin(-5x)$$

Because this result was obtained via eqn (15.5), we again imply the use of radians. ■

Self-test 3

Differentiate each of the following trigonometric functions with respect to x:

3.1 $y = \sin 4x$

3.2 $y = 4\sin 3x$

3.3 $y = -12\sin(8.1x)$

3.4 $y = \cos(44x)$

3.5 $y = 6.3\cos(-7.8\,x)$

3.6 $y = d\cos(d\,x)$

3.7 $y = \sin x + \cos x$

3.8 $y = \left(\dfrac{\cos x}{2}\right) - \left(\dfrac{\sin 3x}{4}\right)$

A full list of the simple standard differentials is given in Table 15.1.

■ **Worked Example 15.8**

What is the rate of change of the function $y = \sin 3x$ at an angle of $45°$?

Methodology:

(i)   Before we start, we must convert $45°$ into radians, otherwise we cannot employ eqns (15.5) or (15.6). From Fig. 13.6 in Chapter 13, we can quickly obtain an angle of $\pi/4$.

(ii) By 'rate of change' we mean the gradient of the graph of $\sin 3x$ (as y) against angle (as x).

(iii) We differentiate the function, and insert $\pi/4$ for x, and calculate its value. The domain is $3x$, which has a value of $3 \times \pi/4 = 0.75\pi$.

The function is:

$$y = \sin 3x$$

Table 15.1 Summary of standard differentials.

Function	Differential
constant	0
ax^n	$anx^{(n-1)}$
e^x	e^x
e^{ax}	$a\,e^{ax}$
$\ln x$	$1/x$
$\ln ax$	$1/x$
$b \ln ax$	b/x
$\sin x$	$\cos x$
$\sin ax$	$a \cos ax$
$\cos x$	$-\sin x$
$\cos ax$	$-a \sin ax$

Therefore we obtain from eqn (15.5):

$$\frac{dy}{dx} = 3 \cos 3x$$

We next insert values:

$$\frac{dy}{dx} = 3 \times \cos(0.75\pi)$$

The value of $\cos(0.75\pi)$ is -0.7076. Accordingly:

$$\frac{dy}{dx} = 3 \times (-0.707) = -2.12$$

so the gradient on the graph $y = \sin 3x$ has a value of -2.12 at $45°$. ∎

Self-test 4

Evaluate the differential coefficient for each of the following:

4.1 $y = 4 \cos 5x$, at an angle of $\pi/6$ rad

4.2 $y = 2 \sin 7x$, at an angle of $\pi/8$ rad

Remember, the value of x in each case must be in radians, not degrees.

Additional problems

15.1 In a kinetics experiment, the rate constant of reaction k varies with temperature T according to the linear form of the **Arrhenius equation**:

$$k = A \exp\left(-\frac{E_a}{RT}\right)$$

where A, E_a, and c are constants. Write an expression describing the temperature dependence of k, i.e. dk/dT.

15.2 The **Bragg equation** relates the angle of diffraction θ with the interplane distance d in a regular crystal:

$$n\lambda = 2d \sin \theta$$

where n is an integer and λ the wavelength of the diffracted light. Write an expression for $d\lambda/d\theta$, assuming n and d are constant.

15.3   (Following on from question 15.2): what is the rate of change when $d = 190$ pm, $n = 1$, and $\theta = \pi/4$ radians?

15.4   When a liquid and gas co-exist at equilibrium, the values of external pressure p and temperature T follow the linear form of the Clausius–Clapeyron equation:

$$p = p_0 \, \exp\left(-\frac{\Delta H_{(vap)}^{\ominus}}{RT}\right)$$

where p_0, R, and $\Delta H_{(vap)}^{\ominus}$ are each constants. Write an expression for dp/dT.

15.5   During a first-order chemical reaction, the concentration of reactant $[A]_t$ decreases with time t according to the first-order integrated rate equation:

$$[A]_t = [A]_0 \, \exp(k t)$$

where $[A]_0$ and k are constants. Write an expression for $d[A]_t/dt$.

15.6   In quantum mechanics, the wavefunction Ψ for an electron in a molecular orbital can often be written in the general form:

$$\Psi = A \, \sin(kx) + B \, \cos(kx)$$

where x is a distance and k is a constant. Write an expression for $d\Psi/dx$.

15.7   When an ideal gas changes volume V, its entropy increases by an amount ΔS:

$$\Delta S = nR \, \ln V + c$$

where n, R, and c are constant. Write an expression for the rate of change of entropy with volume, i.e. write an expression for $d\Delta S/dV$.

15.8   The van't Hoff isochore defines the temperature T dependence of equilibrium constants K:

$$K = c \, \exp\left(-\frac{\Delta H^{\ominus}}{RT}\right)$$

where c and all other terms are constants. What is the rate at which K changes with temperature? That is, write an expression for dK/dT.

15.9   Using an orbital constructed using Slater-type functions, the wavefunction ψ of a 1s orbital can be simplified to:

$$\psi = \left(\frac{\zeta^3}{\pi}\right)^{1/2} \exp(-\zeta r)$$

where r is the distance between an electron and the nucleus, and other terms are constants. Write an expression for $d\psi/dr$.

15.10 The change in entropy of an ideal gas can be written in the form:

$$\Delta S = C_V \, \ln T + R \, \ln V$$

Write an expression for $d\Delta S/dT$. [Hint: C_V is the heat capacity *at constant volume* so the second term is a constant.]

16

Differentiation III

*Differentiating functions of functions:
the chain rule*

Functions of functions

By the end of this chapter, you will know:

- How to recognize a function of a function.
- A function of a function cannot be differentiated straightforwardly, but can be differentiated with the chain rule.

We need the **chain rule** if we have a function of a function. The **chain rule** tells us to:

1. Identify the functions.

2. Differentiate each in turn.

We therefore need to start each example by identifying the two functions involved, one of which operates on the other. This feature of the problem explains the phrase 'function of a function'. The first function acts as the **domain** of the second. For example, if we say $y = \sin(x^2)$, the function x^2 is the domain of the second function, sin.

....

The notation of 'f ( )' in these two equations means 'function of ...'.

....

Having identified the two functions, we separate them in order to use the chain rule. We substitute, saying the domain of the second function is a pseudo variable, call it u. u here is itself the function in the first expression. We will obtain two equations, $y = f(u)$ and $u = f'(x)$, where we included the prime (i.e. the symbol $'$) merely to emphasize how the two functions differ.

Next comes the clever part: we write 'the magic line', which lets us differentiate each function in turn:

$$\left(\frac{dy}{dx}\right) = \left(\frac{dy}{du}\right) \times \left(\frac{du}{dx}\right) \tag{16.1}$$

It is usual to omit the brackets here since they serve no useful purpose except to clarify the expression. Having differentiated each in turn, we reassemble the equations.

■ **Worked Example 16.1**

Differentiate $y = (x^3 + x^2 + 1)^4$.

We could differentiate this function by first multiplying the bracket out, as $(x^3 + x^2 + 1) \times (x^3 + x^2 + 1) \times (x^3 + x^2 + 1) \times (x^3 + x^2 + 1)$, then differentiating each term by rule. And if we were careful, we would obtain the correct answer. Such an approach would, however, be time consuming and fiddly. We want a simpler method, so we will use the chain rule.

We start this example by identifying the two functions, one of which operates on the other, hence the phrase 'function of a function'. The use of a fourth power is clearly a function. It operates on the domain $x^3 + x^2 + 1$, which is itself a function.

We separate the two functions in order to use the chain rule: we substitute, saying the domain is u:

We say : $y = u^4$ and $u = x^3 + x^2 + 1$

We differentiate each one:

$$\text{if} \qquad y = u^4 \qquad\qquad \text{then} \quad \left(\frac{dy}{du}\right) = 4u^3$$

$$\text{and if} \quad u = x^3 + x^2 + 1 \quad \text{then} \quad \left(\frac{du}{dx}\right) = 3x^2 + 2x$$

We then use the magic line from eqn (16.1) and combine these two differentials:

$$\left(\frac{dy}{dx}\right) = 4u^3 \times (3x^2 + 2x)$$

This expression is correct but not useful until we back-substitute, replacing the term u with its original domain of $(x^3 + x^2 + 1)$. Therefore:

$$\left(\frac{dy}{dx}\right) = 4(x^3 + x^2 + 1)^3 \times (3x^2 + 2x) \ \blacksquare$$

> We usually omit the central multiplication sign $\times$, and write '$4(x^3 + x^2 + 1)^3 (3x^2 + 2x)$'.

■ **Worked Example 16.2**

What is the differential of the equation $y = \ln(x^3 + 2)$?

We first identify the functions. It is clear after inspecting the equation that we have a logarithm for which the domain is $(x^3 + 2)$. We next note how this domain contains a simple polynomial function.

It is usual practice to substitute for the domain, so we rewrite the equation, saying:

$$y = \ln u \quad \text{and} \quad u = x^3 + 2$$

We can differentiate each of these expressions in turn without any great difficulty:

$$\text{if} \qquad y = \ln u \qquad \text{then} \qquad \left(\frac{dy}{du}\right) = \frac{1}{u}$$

$$\text{and if} \quad u = x^3 + 2 \quad \text{then} \quad \left(\frac{du}{dx}\right) = 3x^2$$

Then, using the magic line from eqn (16.1):

$$\left(\frac{dy}{dx}\right) = \frac{1}{u} \times 3x^2$$

Finally, we back-substitute for u:

$$\left(\frac{dy}{dx}\right) = \frac{1}{x^3 + 2} \times (3x^2)$$

This equation is completely correct, but we would normally rewrite it slightly by placing the term $3x^2$ on top of the fraction:

$$y = \frac{3x^2}{x^3 + 2} \ \blacksquare$$

We need to note:

- When we employ the chain rule on a logarithmic function, the contents of the domain u remain absolutely unchanged in every particular.

- We always find the unchanged domain positioned as the denominator (bottom line) of a reciprocal (i.e. a fraction of the form $1 \div$ domain).

- It is sometimes possible to simplify the resultant differential further by cancelling and/or factorizing, so the domain may change in appearance in subsequent lines of the problem.

■ **Worked Example 16.3**

Differentiate the equation $y = \sin(3x^4 + 5x)$.

We cannot yet differentiate this expression. It is too difficult. We need the chain rule.

We first identify the functions. It is clear after inspecting the equation that we have a sine, for which the domain is $(3x^4 + 5x)$. We again note how the domain of the *2nd* function, the sine, contains a simple polynomial function.

We substitute for the domain of the sine, thereby rewriting the equation, and say:

$$y = \sin u \quad \text{and} \quad u = 3x^4 + 5x$$

We then differentiate each of the expressions:

$$\text{if} \quad y = \sin u \quad \text{then} \quad \left(\frac{dy}{du}\right) = \cos u$$

$$\text{and if} \quad u = 3x^4 + 5x \quad \text{then} \quad \left(\frac{du}{dx}\right) = 12x^3 + 5$$

Then, using the magic line from eqn (16.1), we combine the two expression to obtain:

$$\left(\frac{dy}{dx}\right) = \cos u \times (12x^3 + 5)$$

This expression is incomplete since it still contains u. We must back-substitute for u:

$$\left(\frac{dy}{dx}\right) = (12x^3 + 5)\cos(3x^4 + 5x)$$

Notice how we have swapped the order here: we placed the bracket *before* the cosine. We did this to avoid any ambiguity. If we had written $y = \cos(3x^4 + 5x)\ (12x^3 + 5)$, some people might have thought the equation meant the cosine of all of $(3x^4 + 5x) \times (12x^3 + 5)$, and obtain a different, *wrong* result. ■

We need to note:

- Differentiating a sine or cosine using the chain rule *always* results in the domain remaining absolutely unchanged.
- By convention, if we have a trigonometric function and a polynomial function, we write the trigonometric function last. This convention is designed to avoid ambiguity.

■ **Worked Example 16.4**

What is the differential of the equation $y = \cos^3 x$?

Before we can start this problem, we need to learn a new notation: the algebraic phrase $\cos^3 x$ means $(\cos x)^3$. The reasons for this notation style lie buried in the history books. We merely need to learn it. Accordingly, we rewrite the question, saying, 'What is the differential of the equation, $y = (\cos x)^3$?'

We next identify the functions. It is clear after inspecting the equation that we have a cosine with the domain x. This function is itself cubed.

In this example, the function $\cos x$ is the domain of the cube. We rewrite the equation, saying:

$$y = u^3 \quad \text{and} \quad u = \cos x$$

We can differentiate each of these expressions without difficulty:

$$\text{if} \qquad y = u^3 \qquad \text{then} \qquad \left(\frac{dy}{du}\right) = 3u^2$$

$$\text{and if} \quad u = \cos x \quad \text{then} \qquad \left(\frac{du}{dx}\right) = -\sin x$$

Then, using the magic line from eqn (16.1):

$$\left(\frac{dy}{dx}\right) = 3u^2 \times (-\sin x)$$

We then back-substitute for u:

$$\left(\frac{dy}{dx}\right) = 3(\cos x)^2 \times (-\sin x)$$

It is neater to write the final answer as:

$$\left(\frac{dy}{dx}\right) = -3 \sin x (\cos x)^2$$

We again rearrange slightly, placing the $\sin x$ before the bracket, as a buttress to avoid ambiguity. We could have rewritten the answer in terms of the old-fashioned notation, saying, $-3 \sin x \cos^2 x$. ∎

■ Worked Example 16.5

In electrochemistry, the electrode potential of the proton–hydrogen redox couple E_{H^+, H_2} is a function of the concentration of the proton, according to the **Nernst** equation:

$$E_{H^+, H_2} = E^{\ominus}_{H^+, H_2} + \frac{RT}{2F} \ln([H^+]^2)$$

where R, T, and F are all constants. The standard electrode potential, $E^{\ominus}_{H^+, H_2}$, is also a constant.

The equation also assumes a standard pressure of hydrogen gas, i.e. $p^{\ominus}$.

We will start by simplifying the *appearance* of the equation, saying:

$$E = E^{\ominus} + k \ln(H^2)$$

where $k = RT \div 2F$ (and 'H' is really $[H^+]$).

We can now analyse the equation, and see the functions are a logarithm and a square. We can then use the chain rule: we will say $E = E^{\ominus} + k \ln u$ and $u = H^2$.

We can differentiate each of these expressions without difficulty:

$$\text{if} \qquad E = E^{\ominus} + k \ln u \quad \text{then} \quad \left(\frac{dE}{du}\right) = k \times \frac{1}{u}$$

$$\text{and if} \quad u = H^2 \qquad\qquad \text{then} \quad \left(\frac{du}{dH}\right) = 2H$$

Then, using the magic line from eqn (16.1):

$$\left(\frac{dE}{dH}\right) = k \times \frac{1}{u} \times 2H$$

After back-substituting for u, we obtain:

$$\left(\frac{dE}{dH}\right) = k \times \frac{1}{H^2} \times 2H$$

which simplifies by cancelling the H on top with one of the two H terms within the square on the bottom, to yield:

$$\left(\frac{dE}{dH}\right) = \frac{2k}{H}$$

Finally, we back-substitute for $k = RT \div 2F$ and reintroduce the standard nomenclature (i.e. 'H' is really $[H^+]$). We say:

$$\left(\frac{dE_{H^+,H_2}}{d[H^+]}\right) = \frac{\cancel{2}RT}{\cancel{2}F[H^+]}$$

We see how the factor of 2 on top and bottom will cancel, so:

$$\left(\frac{dE_{H^+,H_2}}{d[H^+]}\right) = \frac{RT}{F[H^+]}$$

A graph of electrode potential E_{H^+,H_2} (as y) against $[H^+]$ (as x) therefore has a simple gradient of RT/F $[H^+]$. ∎

This example could have been performed without the chain rule if we had used the laws of logarithms (see Chapter 11) after recognizing how the Nernst equation $E = E^{\ominus} + k \ln(H^2)$ could have been written more simply as $E = E^{\ominus} + 2k \ln(H)$.

■ **Worked Example 16.6**

At the heart of quantum mechanics lies a paradox: an object such as a photon can exist both as a wave of wavelength λ and as a particle. The **de Broglie equation** relates the respective values of λ and its mass m:

$$\lambda = \sqrt{\left(\frac{h^2}{2\pi m T k_B}\right)}$$

where T is the absolute temperature, h is the Planck constant, and k_B is the Boltzmann constant.

What is the differential of λ with respect to mass m? ∎

The equation looks very intimidating, so we first simplify it by rewriting:

$$\lambda = \left(\frac{k}{m}\right)^{0.5} \quad \text{where} \quad k = \frac{h^2}{2\pi T k_B}$$

We can now analyse the equation, and see that the functions are a square root and a reciprocal. We can now use the chain rule: we will say $\lambda = u^{0.5}$ and $u = km^{-1}$.

We can differentiate each of these expressions without difficulty:

$$\text{if} \quad \lambda = u^{0.5} \quad \text{then} \quad \left(\frac{dI}{du}\right) = 0.5u^{-0.5}$$

$$\text{and if} \quad u = km^{-1} \quad \text{then} \quad \left(\frac{du}{dT}\right) = -km^{-2}$$

Then, using the magic line from eqn (16.1):

$$\left(\frac{d\lambda}{dm}\right) = 0.5u^{-0.5} \times (-km^{-2})$$

After back-substituting for u, we obtain:

$$\left(\frac{d\lambda}{dm}\right) = 0.5\left(\frac{k}{m}\right)^{-0.5} \times (-km^{-2})$$

We can simplify the expression further:

$$(-km^{-2}) = \left(-\frac{k}{m^2}\right) \text{ and } \left(\frac{k}{m}\right)^{-0.5} = \left(\frac{m}{k}\right)^{0.5}$$

so:

$$\left(\frac{d\lambda}{dm}\right) = 0.5\left(\frac{m}{k}\right)^{0.5} \times \left(-\frac{k}{m^2}\right) = -k^{0.5} \times \frac{1}{2m^{3/2}}$$

Finally, we substitute again for k:

$$\left(\frac{d\lambda}{dm}\right) = \left(\frac{h^2}{2\pi T k_B}\right)^{0.5} \frac{1}{2m^{3/2}}$$

This answer is correct even if we do not tidy it up in any way. We will tidy it up, however, because it would otherwise look rather messy.

Self-test 1

Differentiate each of the following using the chain rule:

1.1 $y = (x^2 + 2)^2$

1.2 $y = (\ln x)^2$

1.3 $y = \sin(e^{3x})$

1.4 $y = e^{(x^4)}$

1.5 $y = \ln(x^3)$

1.6 $y = \sin^5 x$

1.7 $y = \cos(x^4 - x^2)$

1.8 $y = 3\ln(\sqrt{x})$

1.9 $y = \ln(3x^{3/2})$

1.10 $y = \exp(3/x^2 - 2/x^3)$

Additional problems

16.1 When electrochemists employ the **ac impedance** technique, they apply a time-dependent voltage V across the sample and measure the resulting current. The magnitude of the voltage V alters periodically in a sinusoidal way, so:

$$V = \sin \omega t$$

where t is time and ω is the frequency of the ac signal. Use the chain rule to obtain an expression for dV/dt.

16.2 Light is *scattered* as it passes through a solution containing microscopically small particles. Milk is a good example, for it contains small spheres of insoluble oils and fats. The intensity of the scattered light I relates to the angle θ between the plane of polarization and the incident beam:

$$I = I_o \frac{\pi \alpha^2}{\varepsilon_r^2 \lambda^4 r^2} \sin^2 \theta$$

Write an expression for $dI/d\theta$, assuming all the other terms remain constant.

16.3 The acronym **laser** stands for 'light amplification by stimulated emission of radiation'. The emitted light should be monochromatic, meaning it has a single wavelength λ. Radiation of a particular frequency is emitted as a consequence of the dipoles in the host solid oscillating at a constant frequency of ω. But the emission behaviour can be complicated by so-called non-linear effects, in which case the electric field $\mathcal{E}$ of the incident light follows the expression:

$$\mathcal{E} = \sqrt{\frac{1}{2}\mathcal{E}_0^2(1 + \cos 2\omega t)}$$

where t is the time during an oscillation cycle. All other terms are constant. Write an expression for $d\mathcal{E}/dt$.

16.4   Magic-angle NMR was described in Worked example 13.5, in which the tube containing a sample is spun fast while tilted at an angle of θ to the spectrometer's magnetic field (rather than perpendicular to it). The extent of the distortion in the NMR spectrum d relates to θ according to the equation:

$$d = 3\cos^2\theta - 1$$

Write an expression for $dd/d\theta$.

16.5   The Bragg equation relates the angle of diffraction θ with the interplane distance d in a regular crystal:

$$n\lambda = 2d\sin\theta$$

where n is an integer and λ the wavelength of the diffracted light. The usual practice is to vary θ and look at d. Assume therefore that λ is constant. Rearrange the expression to make d the subject, then differentiate to obtain $dd/d\theta$.

16.6   The van't Hoff isochore defines the temperature T dependence of equilibrium constants K:

$$K = c\exp\left(-\frac{\Delta H^{\ominus}}{RT}\right)$$

where c and all other terms are constants. Rather than using eqn (15.1), use the chain rule to obtain an expression for dK/dT.

16.7   Using an orbital constructed using Gaussian functions, the wavefunction ψ of a 1s orbital can be simplified to:

$$\psi = \left(\frac{2\alpha}{\pi}\right)^{3/4}\exp(-\alpha r^2)$$

where r is the distance between an electron and the nucleus, and other terms are constants. Write an expression for $d\psi/dr$.

16.8   The general form of a wavefunction ψ for a free particle moving along the x axis is:

$$\psi = K\cos\left(\frac{x\sqrt{2mE}}{h}\right)$$

where E is the energy, and all other terms are constant. Write an expression for $d\psi/dE$.

16.9   The potential U between two atomic nuclei is given by a form of the Morse curve:

$$U = D_e[1 - \exp(-\beta r)]^2$$

where D_e is the bond dissociation energy, r is the deviation from the equilibrium bond separation, and q is a constant. Write an expression for dU/dr.

16.10 When X-rays strike a regular crystal, they diffract by an angle θ. If the X-rays are of wavelength λ, and the interplane distance is a, then:

$$\lambda = \sqrt{\frac{4a^2}{n^2(h^2 + k^2 + l^2)}\sin^2\theta}$$

where h, k, and l, the so-called Miller indices, are constants, and n is also a constant. Write an expression for $d\lambda/d\theta$.

17

Differentiation IV

The product rule and the quotient rule

Products and quotients

By the end of this section, you will:

- Know the meaning of the mathematical terms 'product' and 'quotient'.
- Learn how to distinguish between simple functions, and products or quotients.

We have met the mathematical term 'product' in several previous chapters, e.g. in Chapter 2 when we met the Pi product Π. The word **product** means merely 'multiplied' together. For example, the product of 5 and 2 is 10; the product of a and b is ab.

The word **quotient** is also a precise mathematical term: it means a fraction. The quotient of 15 and 5 is $15 \div 5 = 3$; the quotient of x and x^2 is $x \div x^2 = 1/x$. Incidentally, this last example also illustrates the way we can often simplify a quotient by cancelling.

In this chapter, we explore the ways of differentiating products and quotients; that is, differentiating two functions which have been multiplied together, or one has been divided by the other. We will use respectively the 'product rule' and the 'quotient rule'.

It is easy to tell when we have a quotient, because we see a fraction with two different functions, one on its top and another on its bottom.

It is sometimes not so easy to spot a product, though. The problem lies in the way maths is generally written with the multiplication signs omitted. Even if we mentally superimpose a multiplication sign, we often need to look critically at the expression and ask ourselves, 'Which functions are involved?' It is easy with a little practice, but can take time.

> We have probably met the word quotient in the initials **IQ**, which stands for **intelligence quotient**. The basis of an IQ test is therefore a mathematical fraction.

The rules for distinguishing between simple functions and products

1. If a variable only occurs once in an expression, then we have a single function. For example, in this context, $\sin(x^2)$ and $\ln x^3$ are both single functions.

2. If we see a variable occurring more than once, but each time it appears within the domain of the same function, then again we have a single function. For example, $\exp(x^2 + x^3)$ is a single function.

3. If there are two functions and both are simple polynomials, we should combine them using the laws of powers (see Chapter 10) to form a single function. For example, if $y = x^2 \times 4x^3$, then $y = 4 \times x^{(2+3)} = 4x^5$.

4. If one term is a polynomial, and the other is enclosed within brackets, and if we can easily multiply the terms, we should do so, then differentiate term by term. For example, if the function was $y = 4x^3(3x - 13)$, we would first multiply out the bracket to obtain $y = 12x^4 - 52x^2$, which is easily differentiated term by term.

5. If one term is a polynomial and the other is best treated within the BODMAS scheme as a function OF (e.g. sine, cosine, exponential, or logarithm), then we have a genuine product.

6. If we see more than one of the functions above, we will certainly have a product.

■ Worked Example 17.1

How should we split up the product $y = x^{-5} \sin 4x$?

The question tells us the overall expression is a product, so we are looking for two separate functions. The sine function is written with its domain to its right, so we could have written $y = x^{-5} (\sin 4x)$. Merely by enclosing the entire sine term within brackets, we see how the sine is one function, so the polynomial x^{-5} is therefore the second. In fact, bracketing a function in this way is a good way of identifying the two functions. ■

As is typical, we write the sine function *after* the polynomial.

Self-test 1

Identify whether the following are single functions, products, or quotients.

1.1 $y = x^4 \ln x$ 1.5 $y = \sin x \cos x$

1.2 $y = x^3 / x^5$ 1.6 $y = x^3 \sin 3x^2$

1.3 $y = \sin 4x / \ln x$ 1.7 $y = \ln x \cos 4x$

1.4 $y = x^4\, 4x^5$ 1.8 $y = 4x^5 / \cos 3x$

The product rule

By the end of this section, you will:

- Know the product rule.
- Know how to differentiate a product with the product rule.
- Be aware that the final answer can often be simplified further, particularly when one function involves an exponential.

When we differentiate a product, it is usual to say the two component functions are u and v, i.e. the product is $u \times v$. We differentiate the product (uv) via the template differential in eqn (17.1):

$$d(u \times v) = u \times \left(\frac{dv}{dx}\right) + v \times \left(\frac{du}{dx}\right) \qquad (17.1)$$

The product rule is often written without the multiplication signs and without brackets. Note the symmetry in this expression, which may make it easier to remember.

■ Worked Example 17.2

Differentiate the product $y = x^2 \ln x$.

Strategy:

(i) Because there is generally no multiplication sign between the two functions within the product, we must first identify the two component parts. We usually call them u and v.

(ii) We differentiate both u and v separately, away from the overall equation. We shall call these differentials du/dx and dv/dx, respectively.

(iii) We enter the four terms, u, v, du/dx, and dv/dx, into the product rule in eqn (17.1).

Thus:

(i) In this example, in common with many others, we have a polynomial function together with a second function. In such cases, we write the operator of the second function *after* the polynomial, thereby decreasing the scope for mis-understanding: for example, if we had written the function as $\ln x\, x^2$, it might be mistaken for the different function $\ln (x \times x^2)$ which not only equals $\ln (x^3)$, but is not a product. We avoid this source of confusion by writing the polynomial *before* the operator of the second function. So, $u = x^2$ and $v = \ln x$.

(ii) We differentiate each component function in turn:

$$u = x^2 \qquad \text{so} \qquad \frac{du}{dx} = 2x$$

$$v = \ln x \qquad \text{so} \qquad \frac{dv}{dx} = \left[\frac{1}{x}\right]$$

(iii) Inserting terms into the product rule, eqn (17.1), generates:

$$\frac{dy}{dx} = x^2 \left[\frac{1}{x}\right] + \ln x \times [2x]$$

$$\underset{u}{\uparrow} \quad \underset{\frac{dv}{dx}}{\uparrow} \quad \underset{v}{\uparrow} \quad \underset{\frac{du}{dx}}{\uparrow}$$

As an aid to comprehension, the two derivatives are written within square brackets.

If one of the two functions is a simple polynomial, we can usually simplify the answer further. The first half of the answer, $x^2 \times 1/x$, can be simplified to yield just x.

It is wise to rewrite subsequently the second half of an answer like this in a different order, as $2x \ln x$, to avoid ambiguity (because $\ln x\, 2x$ could be mistaken for $\ln (x \times 2x)$, which simplifies to yield $\ln (2x^2)$). ■

The question sometimes arises, 'What if I mistake a simple function for a product?'

■ **Worked Example 17.3**

Differentiate $y = x^5 \times 4x^2$ using the product rule.

The example here should properly be differentiated as a single function, because $x^5 \times 4x^2$ is the same as $4x^7$; and the differential of $y = 4x^7$ is $28x^6$.

Nevertheless, we shall pretend this example is a product, saying $u = x^5$ and $v = 4x^2$:

$$u = x^5 \quad \text{so} \quad \frac{du}{dx} = 5x^4$$

$$v = 4x^2 \quad \text{so} \quad \frac{dv}{dx} = 8x$$

Inserting terms into the product rule expression in eqn (17.1) yields:

$$\frac{dy}{dx} = x^5 \times [8x] + 4x^2 \times [5x^4]$$

so:

$$\frac{dy}{dx} = 8x^6 + 20x^6$$

that is:

$$\frac{dy}{dx} = 28x^6 \quad \blacksquare$$

In summary, if we mistook this simple pair of functions for a product, and differentiated using the product rule, we obtain the same differential of $28x^6$ that we would have obtained by differentiating normally.

■ **Worked Example 17.4**

The probability of finding an electron in a 1s orbital is Ψ_{1s}, and is obtained by manipulating the expression:

$$\Psi_{1s} = \frac{4}{a_0^3} r^2 \exp\left[-\frac{2r}{a_0}\right]$$

where the atomic diameter a_0 is a constant and the variable r is the distance of an electron from the nucleus. Differentiate this expression with respect to r.

Before we differentiate this product, we must identify which are the two functions. Perhaps it would help if we rewrote the expression as:

$$\Psi_{1s} = kr^2 \exp(k'r)$$

where $k = (4 \div a_0{}^3)$ and $k' = (-2 \div a_0)$. Clearly, then, kr^2 is the first term (which we will call u) and $\exp(k'r)$ is the second (which we will call v).

To differentiate:

$$u = kr^2 \qquad \text{so} \qquad \frac{du}{dr} = 2kr$$

$$v = \exp(k'r) \quad \text{so} \quad \frac{dv}{dr} = k' \exp(k'r)$$

Inserting terms into eqn (17.1) yields:

$$\frac{d\Psi_1 s}{dr} = kr^2 \times [k' \exp(k'r)] + \exp(k'x) \times [2kr]$$

$$\underset{\substack{\uparrow \\ u}}{} \qquad \underset{\substack{\uparrow \\ \frac{dv}{dr}}}{} \qquad \underset{\substack{\uparrow \\ v}}{} \qquad \underset{\substack{\uparrow \\ \frac{du}{dr}}}{}$$

As before, the two derivatives here are written within square brackets to aid comprehension. This answer is correct, but it would not be wise to leave it in this form. We can factorize, which will result in a considerably shorter and simpler-looking expression.

When differentiating any exponential expression of the form $\exp(cr)$ (where c is a constant), the exponential will itself remain intact and unchanged, although there is usually an additional term obtained in consequence of the chain rule. Accordingly, when we differentiate a mathematical product in which one term is an exponential function, this same exponential will appear in the two halves of the answer, e.g. both within the u term and within du/dr. In this example, the term $\exp(k'r)$ does indeed appear twice.

If one term within the product is an exponential function, we will be able to factorize the immediate answer obtained with the product-rule expression in eqn (17.1).

We can therefore factorize the expression:

$$\frac{d\Psi_{1s}}{dr} = \exp(k'r)\{kr^2 \times k' + 2kr\}$$

where the curly brackets { } indicate that we have factorized. We may even wish to simplify further by factorizing out the two k terms within the curly brackets. We can write:

$$\frac{d\psi_{1s}}{dr} = k \exp(k'r)\{k'r^2 + 2r\}$$

Finally, we back-substitute with the values of k and k':

$$\frac{d\psi_{1s}}{dr} = \left(\frac{4}{a_0^3}\right) \exp\left(-\frac{2r}{a_0}\right)\left\{\left(-\frac{2}{a_0}\right)r^2 + 2r\right\}$$

While each of these answers is correct, the last version is simpler on the eye and therefore easier to 'read'. It would also make it easier to insert numbers when determining a value for the differential coefficient. ■

Self-test 2

Differentiate the following products using eqn (17.1). (The answers on p. 291 have been given in two forms: firstly, in the form generated immediately after using eqn (17.1), and secondly following factorizing.)

2.1 $y = 2x^4 \sin 3x$ **2.4** $y = 2x^{-4} \cos 4x$

2.2 $y = \ln 2x \exp 2x$ **2.5** $y = \sin^2 x \cos^2 x$

2.3 $y = x^5 \ln x$ **2.6** $y = \ln x^2 \sin^3 2x$

The quotient rule

By the end of this section, you will:

- Know the quotient rule.
- Know how to differentiate a quotient with the quotient rule.
- Be aware that the final answer can usually be simplified further.

A **quotient** always takes the form of a fraction, with one function divided by another. In a similar way to the 'splitting up' of a product, we will call the function on the top u and the bottom will be called v. We differentiate a quotient using the following template formula:

$$\frac{dy}{dx} = \frac{v\left(\dfrac{du}{dx}\right) - u\left(\dfrac{dv}{dx}\right)}{v^2} \tag{17.2}$$

■ Worked Example 17.5

Differentiate the quotient

$$y = x^3 / \exp(2x).$$

1. We start by saying the component functions are $u = x^3$ and $y = \exp(2x)$.

2. We then differentiate each function in turn:

$$u = x^3 \qquad \frac{du}{dx} = 3x^2$$

$$v = \exp(2x) \qquad \frac{dv}{dx} = 2\exp(2x)$$

3. We next insert terms into the template expression in eqn (17.2):

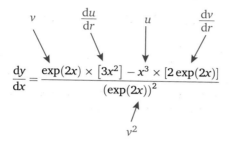

Again, the two derivatives here are written within square brackets to aid comprehension.

4. While this expression is wholly correct, it is long and looks daunting. We can generally simplify such expressions by factorizing. In this example, we first take out a factor of $\exp(2x)$:

$$\frac{dy}{dx} = \frac{\exp(2x)\left[3x^2 - 2x^3\right]}{(\exp(2x))^2}$$

We then cancel the $\exp(2x)$ on the top with one of the two $\exp(2x)$ terms represented by the square on the bottom:

$$\frac{dy}{dx} = \frac{3x^2 - 2x^3}{\exp(2x)}$$

The polynomials on the top row can also be factorized, to read $x^2\,(3 - 2x)$, although this last factorization is not essential. ∎

The question sometimes arises, 'What if I mistake a simple function for a quotient?'

■ **Worked Example 17.6**

Differentiate the function $y = x^3/4x^2$ via the quotient rule.

While this pair of functions are certainly written in the form of a quotient, it is easy to cancel one with the other using the laws of powers (see Chapter 9), to yield $y = \frac{1}{4}x$. The differential of this expression is merely $\frac{1}{4}$. We will, however, pretend it *is* a quotient and differentiate using the quotient rule. We start by saying $u = x^3$ and $v = 4x^2$.

$$u = x^3 \qquad \frac{du}{dx} = 3x^2$$

$$v = 4x^2 \qquad \frac{dv}{dx} = 8x$$

We then insert respective terms into the quotient rule, eqn (17.2):

$$\frac{dy}{dx} = \frac{4x^2 \times [3x^2] - x^3 \times [8x]}{(4x^2)^2}$$

Again, the derivatives here are written within square brackets as an aid to comprehension. We multiply out the terms on the top row, and square the term on the bottom:

$$\frac{dy}{dx} = \frac{12x^4 - 8x^4}{16x^4}$$

The terms on the top row can be collected together:

$$\frac{dy}{dx} = \frac{4x^4}{16x^4}$$

The x^4 terms on the top and bottom cancel, leaving $4 \div 16$, which clearly has a value of $\frac{1}{4}$. ∎

In summary, if we mistook this simple pair of functions for a quotient, and differentiated using the quotient rule, we obtain the same differential of $\frac{1}{4}$ that we would have obtained by differentiating normally.

■ **Worked Example 17.7**

From the Maxwell–Boltzmann theory, the distribution f of energies E is given by the expression:

$$f(E) = \frac{m}{k_B T} E \exp\left(-\frac{mE^2}{2k_B T}\right)$$

where T is temperature, k_B the Boltzmann constant, and m mass.
Differentiate this quotient with respect to T.

At first sight, this function looks like a product. It is more usual to rewrite the expression as a straightforward quotient:

$$f(E) = \frac{Emk_B^{-1}T^{-1}}{\exp(mE^2/2k_B T)}$$

As before, we will rewrite this expression to make it look simpler. This time, we say, $k = Emk_B^{-1}$ and $k' = mE^2 \div 2k_B$. In this way, we obtain the simpler-looking quotient:

$$E = \frac{kT^{-1}}{\exp(k'T^{-1})}$$

Let $u = kT^{-1}$ and $v = \exp(k'T^{-1})$:

$$u = kT^{-1} \qquad \frac{du}{dT} = -kT^{-2}$$

$$v = \exp(k'T^{-1}) \qquad \frac{dv}{dT} = -k'T^{-2}\exp(k'T^{-1})$$

We then substitute the respective terms into eqn (17.2):

$$\frac{df(E)}{dT} = \frac{\exp(k'T^{-1})' \times [-kT^{-2}] - kT^{-1'} \times [-k'T^{-2}\exp(k'T^{-1})]}{(\exp(k'T^{-1}))^2}$$

Once more, the derivatives are written within square brackets to aid comprehension.

In a similar way to Worked example 17.5 above, we can simplify this nasty-looking expression by factorizing and cancelling. Firstly we factorize the top row. Both terms contain $kT^{-2}\exp(k'T^{-1})$, so:

$$\frac{df(E)}{dT} = \frac{kT^{-2}\exp(k'T^{-1})\{-k+k'T^{-1}\}}{(\exp(k'T^{-1}))^2}$$

As before, the curly brackets { } indicate factorizing. Next, the exponential on the top line cancels with one of the exponential terms within the square on the bottom line:

$$\frac{df(E)}{dT} = \frac{kT^{-2}\{-k+k'T^{-1}\}}{\exp(k'T^{-1})}$$

We can now back-substitute for k and k':

$$\frac{df(E)}{dT} = \frac{(Emk_B^{-1})T^{-2}\{-(Emk_B^{-1})+((mE^2/2k_B)T^{-1})\}}{\exp((mE^2/2k_B)T^{-1})} \quad \blacksquare$$

Self-test 3

Differentiate the following products using eqn (17.2). (The answers on page 292 have been given in two forms: firstly, in the form generated immediately after using eqn (17.2), and secondly following factorizing.)

3.1 $y = \dfrac{x^2}{\sin x}$ 3.3 $y = \dfrac{\sin x}{\cos x}$

3.2 $y = \dfrac{\ln x}{x^4}$ 3.4 $y = \dfrac{\exp 2x}{3x^3}$

Additional problems

Questions 17.1–3: Which of the following are products and which are quotients? Explain why.

17.1 The Planck function, G/T

17.2 During a first-order reaction, the rate of reaction follows an equation of the type:

rate $= k[A]$

where k is the rate constant and t is the time.

17.3 The conductivity Λ of an ion through a solution is a function of the mobility μ and the concentration $[C]$:

$\Lambda = \mu[C]$

17.4 All ions in solution are solvated: that is, they are covalently bound to molecules of solvent. And ions will also associate with each other. We say the ion has an 'atmosphere'. The potential ϕ of the ion decreases because of this atmosphere. The modified potential is $\phi_{(atm)}$:

$$f_{(atm)} = Z_i\left(\frac{\exp(-(r/r_D))}{r} - \frac{1}{r}\right)$$

where r is the distance between a test charge and the centre of the ion. Write an expression for $df_{(atm)}/dr$

17.5   Consider the following two-part expression:

$$y = \frac{\sin 2x}{x^3}$$

(i) Differentiate this expression using the quotient rule, with $u = \sin 2x$ and $v = x^3$.

(ii) Differentiate this expression using the product rule, with $u = \sin 2x$ and $v = x^{-3}$.

Demonstrate how the answers to parts (i) and (ii) are the same.

17.6   During the reaction $N_2O_4 = 2NO_2$, the partial pressure of N_2O_4 is given by the expression:

$$p_{(N_2O_4)} = \frac{1 - \zeta}{1 + \zeta}$$

where ζ is the extent of reaction. Write an expression for $dp/d\zeta$

17.7   (Following on from question 17.6): the equilibrium constant K for this reaction is given by the expression:

$$K = \frac{4\zeta^2_{(eq)}}{1 - \zeta^2_{(eq)}}$$

Write an expression for $dK/d\zeta$.

17.8   A polymer comprises a chain of smaller 'monomer' units. Consider a single-strand polymer chain of N monomer units. The length of a single monomer is l. The structure can coil in three-dimensional space, so it will not have a length of $N \times l$. In practice, the probability P that the distance between the ends is nl is given by the expression:

$$P = \left(\frac{2}{PN}\right)^{1/2} \exp\left(-\frac{n^2}{2N}\right)$$

Write an expression for dP/dN.

17.9   Remembering that $\tan \theta$ is the quotient $\sin \theta \div \cos \theta$, obtain an expression for $d(\tan \theta)/d\theta$

17.10   The Arrhenius equation represents an empirical way of relating rate constant k and temperature T. The Eyring equation:

$$k = \frac{k_B T}{h} \exp\left(\frac{\Delta S^{\ddagger}}{R}\right) \exp\left(\frac{\Delta H^{\ddagger}}{RT}\right)$$

represents a theoretical derivation. Only T on the right-hand side is a variable. $\Delta H^{\ddagger}$ is the enthalpy of activation and $\Delta S^{\ddagger}$ is the entropy of activation, and both are constants. Other terms are also constants. Write an expression for dk/dT.

18

Differentiation V

Maxima and minima on graphs: second differentials

Introducing turning points: maxima and minima

By the end of this section, you will know that:

- All graphs of polynomial functions $y = x^n$ have maxima and minima.
- The gradient at a maximum or a minimum momentarily has a value of 0.
- These maxima and minima represent the turning points of the graph: before a maximum, the gradient is positive and afterwards the gradient is negative; before a minimum, the gradient is negative and afterwards the gradient is positive.
- If a graph has an inflection, the gradient at the turning point does not change sign, but merely decreases to zero then increases again

We saw in Chapter 8 how a straight line follows an equation of the type $y = mx + c$, where m is the gradient and c is the intercept. The value of m is constant because the line is linear. But consider the example function $y = x^2$, a graph of which is shown in Fig. 18.1. The most obvious feature of the graph is its curvature. It is not linear, so the gradient continually changes. In fact, at no time is the gradient constant.

Simple differentiation of this function tells us the gradient. The differential $dy/dx = 2x$, so the gradient at $x = 2$ is 4; the gradient at $x = -6$ is -12. And we notice an unusual feature: the gradient of $y = x^2$ has a value of 0 when x is 0, i.e. when the line passes through the origin.

After a moment's thought, we may see a trend emerging. The graph has a bell shape. In technical terms, its shape is a parabola. The two arms are fairly steep, and become progressively steeper as x becomes either more positive or more negative. Between these two extremes, the curve shows a minimum. Before the minimum, the gradient is negative: the curve travels 'downhill' as we move from left to right. Conversely, the gradient is positive after the minimum, and starts its journey 'uphill'. We say the minimum represents a turning point, because the sign of the gradient turns from negative to positive.

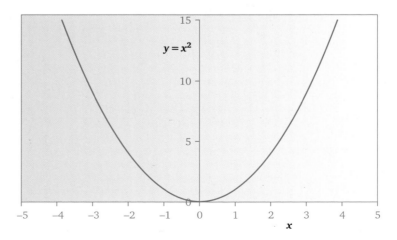

Figure 18.1 The function $y = x^2$ has a minimum at (0,0).

By looking at this very simple example, we see how the gradient changes sign either side of the minimum. Logic dictates that if the sign changes smoothly as we move from left to right (it was negative and becomes positive), then the gradient must transiently have a value of 0. And, as we saw above, this occurs when $x = 0$.

Now look at Fig. 18.2, which shows a graph of the related function $y = -x^2$. This graph also has a turning point, which this time is a **maximum**. And being a maximum rather than a minimum, the way the gradient changes its sign differs from that in Fig. 18.1. The gradient is positive before the turning point and negative afterwards, i.e. follows the opposite trend to that seen before. Again, the gradient at the turning point has a value of 0.

Figure 18.3 shows another graph, and depicts the function $y = x^3$. This graph also has a turning point, which again occurs at $x = 0$. Even a quick look at the gradients, though, shows how this third graph differs from those in Figs 18.1 and 18.2. It shows neither a maximum nor a minimum: the gradient is positive on both the left-hand side of the graph and its right-hand side. At no time is the gradient negative. We say the

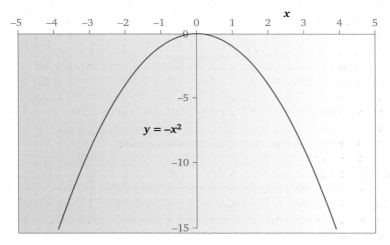

Figure 18.2 The function $y = -x^2$ has a **maximum** at (0,0).

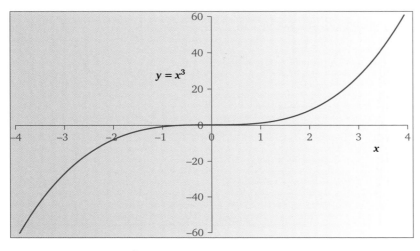

Figure 18.3 The function $y = x^3$ has an **inflection** at (0,0).

graph shows a **point of inflection**. Points of inflection need not have a gradient of zero.

These changes in gradient are summarized in Table 18.1.

Table 18.1 Analysis of the gradients either side of a turning point.

Type of turning point	Gradient before turning point	Gradient after turning point	Examples*
Maximum	positive	negative	$y = -x^2$ $\quad y = -x^4$ $\quad y = -x^6$
Minimum	negative	positive	$y = x^2$ $\quad y = x^4$ $\quad y = -x^6$
Point of inflection	(i) positive (ii) negative	positive negative	$y = x^3$ $\quad y = x^5$ $\quad y = x^7$ $y = -x^3$ $\quad y = -x^5$ $\quad y = -x^7$

* Notice how graphs of even-numbered powers display a maximum or minimum, and graphs of odd-numbered powers yield an inflection.

Determining the coordinates of a turning point

By the end of this section, you will know that:

- To determine a turning point, we first differentiate the equation of the function and equate its differential to zero. Subsequent algebraic manipulation allows for determination of x at the turning point.
- When a graph has two turning points, we will usually have to solve a quadratic equation in x to determine the turning points.

The presence of a minimum in Fig. 18.1 and a maximum in Fig. 18.2 is no accident: all graphs of polynomial functions possess turning points and/or inflections. Polynomial expressions involving a single term will possess only one turning point or inflection. This explains why the graph of $y = x^3$ in Fig. 18.3 has a single inflection. But then look at Fig. 18.4, which depicts the more complicated function

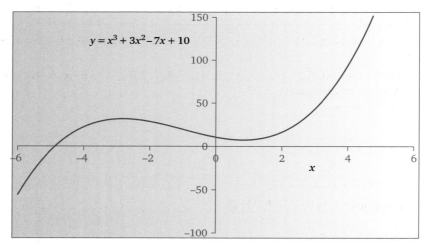

Figure 18.4 The function $y = x^3 + 3x^2 - 7x + 10$ has two turning points, both a **local maximum** at the point $(-3, 27)$ and a **local minimum** at $(1, 7)$.

$y = x^3 + 3x^2 - 7x + 10$: this graph has *two* turning points. In fact, the general rule says graphs of general polynomials, $y = x^n + x^{(n-1)} + \dots$, will have $(n-1)$ turning points. Therefore a quadratic equation, $y = ax^2 + bx + c$, will always have one turning point.

While the graph in Fig. 18.4 has a maximum at the point $(-3, 31)$, the value of y is not the largest value on the page. The value of y increases considerably higher at the far right-hand side of the graph. For this reason, we say the point $(-3, 31)$ is a **local maximum**. For the same reason, the point $(1, 7)$ is a **local minimum**.

Self-test 1

How many turning points should we expect in each of the following graphs?

1.1 $y = x^4$

1.2 $y = x^2 + x + 2$

1.3 $y = x^4 + 3x^3 + 12$

1.4 $y = x^5 + x^4 - 3x^3 + 6x^2 - 8$

■ **Worked Example 18.1**

This graph of the curve $y = x^2 + x - 4$ has a single turning point. What are its coordinates?

Strategy:

(i)   At the turning point, the gradient has a value of 0.

(ii)   We differentiate the function, and equate the differential coefficient to zero.

(iii)   We rearrange this equation to make x the subject. We say this value of x is the x-coordinate of the turning point.

(iv)   Knowing x at the turning point, we back-substitute into the original equation to determine the value of y at the turning point.

Thus:

(i)   The differential of the function $y = x^2 + x - 4$ is:
$$\frac{dy}{dx} = 2x + 1$$

(ii)   We note how the gradient at a turning point is zero, and say $dy/dx = 0$:
$$2x + 1 = 0$$

(iii)   If $2x + 1 = 0$, then $2x = -1$, so $x = -\frac{1}{2}$. We calculate the value of x at the turning point as $-\frac{1}{2}$.

(iv)   Inserting this value of x into the original function allows us to compute the value of y at the turning point:
$$y = x^2 + x - 4$$

so:
$$y = \left(\tfrac{1}{2}\right)^2 - \tfrac{1}{2} - 4$$

and:
$$y = 0.25 - 0.5 - 4 = -4.25$$

The turning point is at $(0.5, -4.25)$. ■

■ **Worked Example 18.2**

Two atoms are joined by a single bond. The potential energy of the atoms depends on their separation according to Fig. 18.5, which is known as a

$r_{(eq)}$ **Separation between atoms, r**

Figure 18.5 Atoms bonded together have a preferred bond length $r_{(eq)}$, which corresponds to the length at the minimum of a Morse curve of energy (as y) against interatomic separation r (as x).

....
A more realistic equation to describe a Morse curve takes the form:

$E = \varepsilon[\{1-$
$\exp(\beta(r - r_{(eq)}))^2 - 1\}]$

where β and ε are constants, and $r_{(eq)}$ is the equilibrium bond length.
....

Morse curve. A greatly simplified equation for the graph is:

$$E = 2r^2 - 4r + 3$$

where r is the normalized bond length. Determine the minimum on this graph. ■

We first differentiate the expression, saying:

$$\frac{dE}{dr} = 4r - 4$$

At the minimum, the gradient of the curve is zero, so:

$$0 = 4r - 4$$

Slight rearranging yields:

$$4r = 4$$

so:

$$r = 1$$

Knowing the value of r at the minimum, we calculate the corresponding value of E:

$$E = 2 \times (1)^2 - 4 \times (1) + 3$$

so:

$$E = 1$$

That is, the minimum of the graph is the point (1,1).

In summary, we have determined the minimum of the Morse curve. This minimum corresponds to the preferred separation between the atoms. In other words, the preferred bond length $r_{(eq)}$ corresponds to the minimum energy in a Morse curve.

Self-test 2

Determine the coordinates of the minimum or maximum for each of the following functions:

2.1 $y = x^2 + x + 5$ 2.3 $y = -5x^2 + 4x + 2$
2.2 $y = 3x^2 - 12x - 7$ 2.4 $y = 3.2x^2 - 9.1x$

When a graph has more than one root, the differential has the form of a quadratic equation. Solution of this quadratic then yields two roots, which identify the two turning points. Again, determine the coordinates of the minimum or maximum for each of the following functions:

2.5 $y = 1/3x^3 + 5x^2 + 24x + 7$ 2.6 $y = x^3 + 9x^2 + 24x$

Is the turning point a maximum, a minimum, or an inflection?

By the end of this section, you will know:

- We can discern whether a turning point is a maximum or minimum by considering the gradients near a turning point.
- The double differentiation method allows us to quantify the rate at which the gradient changes. We differentiate the differential: the differential coefficient of a maximum is negative and the differential coefficient of a minimum is positive.

Look again at the simple function $y = x^2$ depicted in Fig. 18.1. We have seen already how the turning point occurs at the origin, (0,0). We saw briefly in the introduction how the gradient changes sign at a turning point. We now analyse ways in which this change of sign can be used critically, to allow us to determine whether a turning point is a maximum, minimum, or inflection.

The simplest way to tell the nature of a turning point is inspection of a drawn curve. Such an approach is time consuming but otherwise reliable.

The method of second differentiation

We have consistently emphasized how a differential represents a *rate of change*. We now extend the idea to look at the way the gradients of these graphs itself shows a trend—in effect, we analyse how the rate of change itself changes.

As we have seen, the best way to determine the rate at which a function changes is to differentiate. The resultant differential yields directly the gradient of the function, dy/dx. Therefore, if we want to know how fast a gradient changes, we differentiate the differential. We obtain 'the second differential':

The first differential of a function y yields the gradient: $\dfrac{dy}{dx}$

The second differential of y yields the rate of change of the gradient:

$$\frac{d(dy/dx)}{dx} = \frac{d^2y}{dx^2}$$

When we describe a double differential, we say aloud, 'dee two why by dee ex squared'.

The reason we can use this method follows from the trends in the gradients, as summarized in Table 18.2:

- If an observed overall trend is a decrease (positive via zero to negative), as seen for a maximum, then the second differential will be negative.
- If an observed overall trend is an increase (negative via zero to positive), as seen for a minimum, then the second differential will be positive.

Table 18.2 The trends in the signs of a gradient either side of a maximum, minimum, or point of inflection.

Maxima:	The gradient shows the trend: positive → zero → negative, i.e. decreases
Minima:	The gradient shows the trend: negative → zero → positive, i.e. increases
Inflection:	(i) The gradient shows the trend: positive → zero → positive, i.e. no sign change
Inflection:	(ii) The gradient shows the trend: negative → zero → negative, i.e. no sign change

Care: The method of double differentiation to test for inflections does not work for simple polynomial functions: for example, the value of d^2y/dx^2 for $y=x^4$, $y=x^6$, $y=x^8$, etc., will each be 0 at the turning point, although the turning point in each case is not an inflection but a minimum. To determine whether a turning point is an inflection, or a maximum or minimum, may require more advanced methods.

- If the observed overall trend shows no overall change in sign, as seen for an inflection, then the second differential has a value of 0.

These results are summarized in Table 18.3.

Table 18.3 The trends in the signs of a second differential d^2x/dy^2 either side of a maximum, minimum, or point of inflection.

Maximum:	$\dfrac{d^2y}{dx^2}$ is negative
Minimum:	$\dfrac{d^2y}{dx^2}$ is positive
Inflection:	$\dfrac{d^2y}{dx^2}$ has a value of 0

■ Worked Example 18.3

Consider the simple function $y=4x^2+1$. Determine its turning point. Is it a maximum, minimum, or an inflection?

The gradient of this function is given by the *first* differential: $dy/dx = 8x$.

The turning point occurs when the gradient is zero, so the turning point occurs when $x=0$. Back-substitution says the value of y is 1. Therefore, the turning point is $(0,1)$.

The rate of change of this gradient is given by the *second* differential: $d^2y/dx^2 = 8$.

The second differential is always positive (in this case, regardless of the value of x), so this turning point is a minimum. ■

■ Worked Example 18.4

Consider the simple function $y=4x^5$. First determine its turning point, then decide if it is a maximum, minimum, or an inflection.

The gradient of this function is given by the *first* differential: $dy/dx = 20x^4$.

The turning point occurs when the gradient is zero, so the turning point occurs when $x=0$. Back-substitution says the value of y is also 0. Therefore, the turning point is $(0,0)$.

The rate of change of this gradient is given by the *second* differential: $d^2y/dx^2 = 80x^3$. ■

When we substitute the value $x=0$ into this, the second differential, we obtain the result $d^2y/dx^2 = 0$, so this turning point represents an inflection.

Self-test 3

Determine the turning point for each of the following functions. Differentiate a second time to discern whether the turning point(s) are maxima, minima, or points of inflection.

4.1 $y=x^3+2$ 4.3 $y=4x^3+4x^2+12$

4.2 $y=3x^2+6x$

Additional problems

The majority of chemical examples used to illustrate the concepts of second differentials and turning points are so complicated that the mathematical concepts become lost. These examples are purely mathematical.

18.1–3 These questions relate to the equation $y = x^3 + 4x^2 - 5x$.

18.1 Determine graphically the turning points of the function, and identify which are maxima and which are minima.

18.2 Factorize the equation in order to determine where the function cuts the x axis.

18.3 Determine the coordinates of the turning points using first and second differentials.

18.4 Write the double differential of the equation:

$$y = x^4 - x^3 + 12 \ln x + \frac{1}{x^2}$$

18.5 Write the double differential of the equation $y = (x^3 + x)^5$.

18.6 Write the double differential of the equation $y = \sin x \cos x$.

18.7 Consider the equation $y = x^3 - 4x^2 - 6x$. Differentiate this expression and then factorize this expression to determine the turning points.

18.8 Write the double differential of the equation $y = x^3 + x^2 + 1$, and hence determine the coordinates of its turning point. From the double differential, is the turning point a maximum or a minimum?

18.9 Write the double differential of the equation $y = \sin 12x$, and hence determine the coordinates of its turning point. By looking at the gradient just before and just after the turning point, determine the nature of the turning point.

18.10 Write the double differential of the equation $y = \ln(3x) - x^2$, and hence determine the coordinates of its turning point. Is it a maximum or a minimum?

19

Integration I

Reversing the process of differentiation

Introducing integration: integrating power functions by rule

By the end of this section, you will know:

- The process of integration performs the opposite function to differentiation.
- How to integrate a variety of polynomial functions by rule.

When we differentiate, we start with the equation of a line and obtain an expression to quantify the way it changes. Differential coefficients derived in this way allow us to calculate exactly a rate of change.

But imagine we start with a mathematical expression describing a line's gradient—a differential—yet we want to know about the line itself. In effect we want to perform the reverse process to differentiation. We call this reverse process **integration**.

■ **Worked Example 19.1**

If a line has a gradient of x^3, what is the equation of the line it describes?

When we differentiate a simple polynomial expression using the rule in Chapter 14, we start with a function of the form ax^n. We decrease the power by 1, and multiply the factor a by n, yielding $anx^{(n-1)}$. The function differentiated to obtain a gradient of x^3 must therefore have been x^4. But if we differentiated x^4 we would obtain $4x^3$, not $1 \times x^3$. We see how the original function must have been $\frac{1}{4}x^4$.

In general:

Equation (19.1) cannot be used if $n = -1$. Such a situation will be considered later.

$$\text{if} \quad \frac{dy}{dx} = ax^n \quad \text{then} \quad y = \frac{a\,x^{(n+1)}}{(n+1)} \tag{19.1}$$

■

■ **Worked Example 19.2**

Integrate the expression $dy/dx = 5x^6$.

Before we employ the relationship in eqn (19.1), we must introduce some nomenclature. Firstly, we must decide which term represents the variable. The decision is easy in this example: the variable is clearly x. We say we integrate *with respect to* x. In terms of the symbols we employ, we write this methodology in two distinct ways:

- Firstly, we write an integral sign $\int$, which is simply an old-fashioned script 's', which we call a **script S**. We therefore write: $\int 5x^6$

- Secondly, to indicate the variable of interest is x, we need somehow to write an x after the function to be integrated. By convention, we use the notation dx. In words, we say, 'with respect to x'. We write $\int 5x^6 \, dx$.

We can now begin:

$$\frac{dy}{dx} = 5x^6 \quad \text{so} \quad y = \int 5x^6 \, dx$$

Integrating by rule (with eqn (19.1)) yields:

$$y = \frac{5x^{(6+1)}}{(6+1)} \quad \text{so} \quad y = \frac{5x^7}{7} \quad ■$$

We need to note:

- By convention, we generally do not convert the fraction 5/7 into a decimal, but leave it as a fraction.

- We occasionally tidy up the fraction by cancelling or rearranging.

■ **Worked Example 19.3**

Integrate the equation $1/x^4$ with respect to x.

Before we start, we must rewrite the expression in a more 'user-friendly' manner, to remove the fraction. We write:

$$\frac{dy}{dx} = x^{-4}$$

We can now integrate by rule via eqn (19.1). We say:

$$y = \int x^{-4} \, dx$$

so:

$$y = \frac{x^{-3}}{-3}$$

which we could rewrite more tidily by writing $y = -\dfrac{1}{3x^3}$. ■

■ **Worked Example 19.4**

Integrate the following expression: $\dfrac{dy}{dx} = (1/x^{1.2}) - 12x + 4$.

In this example, for the first time we see a function comprising several terms. In just the same way as we **differentiate** an expression one term at a time, so we can integrate this expression one term at a time.

Methodology:

- It is easiest to rewrite the term $1/x^{1.2}$ as $x^{-1.2}$.

- Before we start we need to note how the final term, 4, can be written as $4x^0$.

$$\frac{dy}{dx} = x^{-1.2} - 12x + 4x^0 \; dx$$

$$\downarrow \qquad \downarrow \qquad \downarrow$$

$$y = \int \frac{dy}{dx} = \frac{x^{-0.2}}{-0.2} - \frac{12x^2}{2} + \frac{4x^1}{1}$$

> Anything raised to the power of 0 has a value of 1, so $4x^0$ is actually 4×1, i.e. 4.

so:

$$y = -5x^{-0.2} - 6x^2 + 4x \quad ■$$

Self-test 1

Integrate the following expressions by rule using eqn (19.1):

1.1 $\dfrac{dy}{dx} = x^3$ 1.6 $\dfrac{dy}{dx} = \dfrac{1}{x^{4.2}} + 5$

1.2 $\dfrac{dy}{dx} = 2x^5$ 1.7 $\dfrac{dy}{dx} = x^2 + 3x - 6$

1.3 $\dfrac{dy}{dx} = x^{1.3}$ 1.8 $\dfrac{dy}{dx} = \dfrac{4}{x^5} - \dfrac{3.5}{x^2}$

1.4 $\dfrac{dy}{dx} = 3.7x^{8.2}$ 1.9 $\dfrac{dy}{dx} = \sqrt{x^7}$

1.5 $\dfrac{dy}{dx} = \dfrac{6}{x^3}$ 1.10 $\dfrac{dy}{dx} = \sqrt{x^6} - 2x^2$

Integrating functions other than simple powers

By the end of this section, you will know:

- How to integrate other functions by rule.
- The domain of a logarithm, exponential, sine, and cosine function remains unchanged following integration.

Just as we can differentiate functions other than polynomials, so we can integrate these same other functions. And we employ the same types of methodology.

■ **Worked Example 19.5**

What is the integral of e^{9x}?

The *differential* of an exponential is obtained by rewriting the exponential, without altering the domain in any way. We therefore start the integration of e^{9x} by writing just e^{9x}. We next add an additional factor, which is the differential of the domain. If we had differentiated e^{9x}, we would generate $9 \times e^{9x}$. To accommodate the additional factor of a, we need to divide the integral by 9.

The integral of e^{9x} is therefore $\frac{1}{9}e^{9x}$. ■

The general rule for integrating exponentials is:

$$\frac{dy}{dx} = e^{ax} \quad \text{and} \quad y = \frac{1}{a}e^{ax} \tag{19.2}$$

■ **Worked Example 19.6**

Figure 19.1 shows the three p orbitals. In any one direction, we obtain the probability P of finding an electron in a p orbital by integrating the function:

$$f = \exp\left(-\frac{kr}{a_0}\right)$$

In this example, we have written **exp** rather than **e**, because the domain is bulky.

where r is the distance of the electron from the atomic nucleus, k is a constant, and a_0 is the atomic radius. What is integral of f with respect to r?

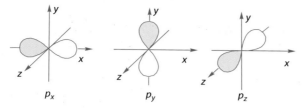

Figure 19.1 The shapes of the three *p* orbitals.

This function fits into the general equation in eqn (19.2) if we rewrite it, saying e^{ax}, where $a = -k/a_0$. We can now insert terms into eqn (19.2):

$$\text{if} \quad f = \exp\left(-\frac{kr}{a_0}\right)dr \quad \text{then} \quad p = \int f \, dr = \left(-\frac{a_0}{k}\right)\exp\left(-\frac{kr}{a_0}\right)$$

Remember:
Dividing anything by a *fraction* gives the same result as multiplying by the reciprocal of that fraction, so dividing by $-k/a_0$ is the same as multiplying by $-a_0/k$.

■ **Worked Example 19.7**

Electrochemists often measure the **ac impedance** of a sample. They apply a voltage *V* across the sample and measure the resulting current. The magnitude of the voltage *V* alters periodically in a sinusoidal way, so:

$$V = \sin \omega t$$

where *t* is time and ω is the frequency of the ac signal. The integral of the voltage yields mechanistic information. What is the integral of *V* with respect to time?

An **impedance** is an electrical resistance that changes periodically with time.

The rules for integrating sines and cosines are given respectively by eqn (19.3) and eqn (19.4). Notice how the domain remains intact in each case.

$$\text{if} \quad \frac{dy}{dx} = \sin ax \quad \text{then} \quad y = -\frac{1}{a}\cos ax \qquad (19.3)$$

$$\text{and if} \quad \frac{dy}{dx} = \cos ax \quad \text{then} \quad y = \frac{1}{a}\sin ax \qquad (19.4)$$

So, after inserting terms into eqn (19.3), we obtain:

$$\text{if} \quad V = \sin \omega t \, dt \quad \text{then} \quad \int V = -\frac{1}{\omega}\cos \omega t$$

A full list of the simple standard integrals is given in Table 19.1. ■

Self-test 2

Integrate each of the following by rule:

2.1 $\dfrac{dy}{dx} = \cos 4x$

2.2 $\dfrac{dy}{dx} = 6\sin 4x$

2.3 $\dfrac{dy}{dx} = 5x^{-3}$

2.4 $\dfrac{dy}{dx} = \dfrac{3}{x}$

2.5 $\dfrac{dy}{dx} = \dfrac{1}{2}e^{5x}$

2.6 $\dfrac{dy}{dx} = 96.4$

2.7 $\dfrac{dy}{dx} = x^4 - \sin 2x$

2.8 $\dfrac{dy}{dx} = e^{dx} - e^{ex}$

2.9 $\dfrac{dy}{dx} = \dfrac{5}{x} + e^{3.2x}$

2.10 $\dfrac{dy}{dx} = b\cos ax - a\sin bx$

Table 19.1 List of standard indefinite integrals.

y	$\int y\ dx$
a	ax
$a\,x^n$	$\dfrac{a\,x^{(n+1)}}{(n+1)}$ but not if $n = -1$
$\sin ax$	$-\dfrac{1}{a}\cos ax$
$\cos ax$	$\dfrac{1}{a}\sin ax$
e^{ax}	$\dfrac{1}{a}e^{ax}$
$\dfrac{1}{x}$	$\ln x$
$\dfrac{a}{x} = a \times \dfrac{1}{x}$	$a \times \ln x$

Constant of integration *c*

By the end of this section, you will know:

- Because a constant disappears when *differentiated*, the *integral* of a function needs a constant of integration, *c*.
- An integral with a constant of integration is called an *indefinite integral*.
- An integral without a constant of integration is called a *definite* integral.

Consider the following three functions: $y = x^2$, $y = x^2 + 3$, and $y = x^2 + 6$ (depicted in Fig. 19.2). The constant at the end of each expression (0, 3, or 6) shifts the curve up or down the page, but does not change the shape of the curve in any way. We now differentiate each one:

$$y = x^2 \qquad \text{so} \qquad \frac{dy}{dx} = 2x$$

$$y = x^2 + 3 \qquad \text{so} \qquad \frac{dy}{dx} = 2x$$

$$y = x^2 + 6 \qquad \text{so} \qquad \frac{dy}{dx} = 2x$$

Because the shape of each curve is identical, the differential of each equation will be the same, and the constant disappears.

The loss of the constants places a severe limitation on the process of integration. If we want to integrate any of the three differentials above, we could not guess that the first had no constant (i.e. its constant was zero). Nor could we have guessed the value of the constant in the second was 3 and 6 in the third.

When we integrate a function, we need to accommodate this inherent uncertainty caused during differentiation. Accordingly, if we integrate $2x$ by rule, we obtain $y = x^2$. And because we know nothing about the constant in the original expression—whether

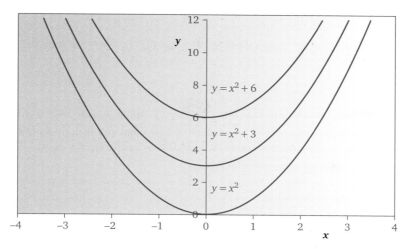

Figure 19.2 The gradients of the three functions $y = x^2$, $y = x^2 + 3$, and $y = x^2 + 6$ are always the same.

it contained a constant at all and, if so, what its value was—we write the integral as $y = x^2 + c$. We call the last term c a **constant of integration**. Unless additional information is available (as below), we do not know the value of c, so the integral is to some extent incomplete. We say such integrals are **indefinite**.

■ **Worked Example 19.8**

Integrate the expression $dy/dx = 1/x$

We obtain the reciprocal function $1/x$ when we differentiate the logarithm $\ln x$, so the final integral must contain $\ln x$ at its core. We also require a constant of integration c.

The rule for integrating a reciprocal is:

$$\text{if} \quad \frac{dy}{dx} = \frac{1}{x} \quad \text{then} \quad y = \ln x + c \tag{19.5}$$

Inserting terms into eqn (19.5) yields:

$$y = \int \frac{1}{x} dx = \ln x + c \quad ■$$

Evaluating the constant of integration: definite integrals

By the end of this section, you will know:

- How to determine a value for the constant of integration c.
- The simplest method of determining c requires that we know the gradient of a line and a single point through which the line passes.

■ Worked Example 19.9

A line of gradient of $3x^2$ passes through the point (1,11). What is the equation of the line?

Methodology:

(i)   We first integrate the equation describing the gradient by rule with eqn (19.1).

(ii) Knowing the equation of the line (which at this stage will include a constant of integration c), we substitute with the coordinates of the known point.

(iii) We then rearrange to make c the subject.

Integrating the equation yields:

$$\frac{dy}{dx} = 3x^2 \quad \text{so} \quad y = \int 3x^2 \, dx = x^3 + c$$

The equation of the line is $y = x^3 + c$.

Next, we substitute the coordinates of the known point, saying $x = 1$ and $y = 11$:

$$11 = 1^3 + c$$

so:

$$c = 10$$

We deduce the equation of the line to be $y = x^3 + 10$, which is depicted in Fig. 19.3. ■

We need to note:

• We call the answer **definite integral** when we know the value of the constant of integration.

• Usual practice tells us to leave the constant of integration as a fraction, without converting it to a decimal.

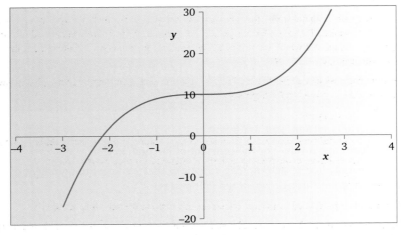

Figure 19.3 The line $y = x^3 + 10$ has the gradient $3x^2$ whatever the value of x.

Self-test 3

Integrate the following differentials by rule using eqns (19.1)–(19.5), and in each case determine the constant of integration:

3.1 $\dfrac{dy}{dx} = x^3$ which goes through the point (0,9).

3.2 $\dfrac{dy}{dx} = e^{3x}$ which goes through the point (1,2).

3.3 $\dfrac{dy}{dx} = \sqrt{x}$ which goes through the point (7,3).

3.4 $\dfrac{dy}{dx} = \dfrac{1}{x}$ which goes through the point (4,5).

3.5 $\dfrac{dy}{dx} = x^3 - x^2$ which goes through the point (3,12).

3.6 $\dfrac{dy}{dx} = \dfrac{3}{x}$ which goes through the point (2,3).

Additional problems

19.1 When an electrode is immersed in a solution of analyte and polarized, the time-dependent current I_t decreases with time t according to the **Cottrell equation**:

$$I_t = nFA[C]\sqrt{\frac{\pi D}{t}}$$

where other terms are constants. Since current is defined as the rate of change of charge Q, i.e. $I = dQ/dt$, integrate the Cottrell equation to derive an expression relating Q and time t.

19.2 The enthalpy change ΔH associated with warming a solid from temperature T_1 to T_2 is a simple function of the heat capacity, C_p:

$$\Delta H = \int C_p \; dT$$

The heat capacity of a solid C_p can often be approximated to a power series in T:

$$C_p = a + bT + \frac{c}{T^2}$$

Integrate this second expression to determine the enthalpy change ΔH when a substance is warmed.

19.3 (Following on from question 19.2): at very low temperatures, the heat capacity C_p of a solid is proportional to T^3, so we write $C_p = aT^3$. Determine the temperature dependence of the enthalpy change ΔH when a substance is warmed at these low temperatures, i.e. integrate the expression, $C_p = aT^3 \; dT$.

19.4 At temperatures just above the **triple point**, the slope of a melting line on a phase diagram for methane is given by the expression:

$$\frac{dp}{dT} = 0.08446 \, T^{0.85}$$

Integrate the right-hand side of this expression to obtain an expression relating p and T.

19.5 The attractive forces U between two particles depend on their separation r:

$$U = -\frac{A}{r^6}$$

where A is a jumble of constants. Integrate this expression with respect to r.

19.6 A line goes through the point (2,20), and has the gradient $x^3 + x^2 - 1$. What is the equation of the line?

19.7   When the reaction of a chemical A proceeds via **first-order kinetics**, the rate of change of $[A]_t$ follows the equation:

$$\frac{d[A]}{dt} = -k[A]$$

Rearranging (including partial integration) yields:

$$-kt = \frac{1}{[A]}\,d[A]$$

Integrate the right-hand side to discern the relationship between t and $[A]_t$ during a first-order reaction.

19.8   When the reaction of a chemical A proceeds via **second-order kinetics**, the rate of change of $[A]_t$ follows the equation:

$$\frac{d[A]}{dt} = -k[A]^2$$

Rearranging (including partial integration) yields:

$$-kt = \frac{1}{[A]^2}\,d[A]$$

Integrate the right-hand side to discern the relationship between t and $[A]_t$ during a second-order reaction.

19.9   When the reaction of a chemical A proceeds via **third-order kinetics**, the rate of change of $[A]_t$ follows the equation:

$$\frac{d[A]}{dt} = -k[A]^3$$

Rearranging (including partial integration) yields:

$$-kt = \frac{1}{[A]^3}\,d[A]$$

Integrate the right-hand side to discern the relationship between t and $[A]_t$ during a third-order reaction.

19.10 A line goes through the point $(3,202)$, and has the gradient $1/x + \exp(2x)$. What is the equation of the line?

20

Integration II

Separating the variables, integration with limits, and area determination

Separating the variables

By the end of this section, you will know:

- It is possible to separate a differential expression into its two variables.
- Having separated the variables, it is common to perform two integration steps.
- It is possible to separate the variables, then integrate using limits.

A differential expression will generally have the form:

$$\frac{dy}{dx} = f(x)$$

where f(x) is a function of *x*. We can treat the differential dy/dx as a fraction, which permits us to cross-multiply by dx. The expression will then look like:

$$dy = f(x)dx$$

We then integrate in the usual way.

■ Worked Example 20.1

If $dy/dx = 4x^4$, separate the variables and then integrate with respect to *x*.

We first separate the variables, i.e. cross-multiply by dx:

$$dy = 4x^4 dx$$

We then integrate in the usual way:

$$\int dy = \int 4x^4 \, dx$$

This equation looks odd: there is no function on the left-hand side. We will employ a dodge, saying that dy is the same as $1 \times dy$. We write:

$$\int 1 \, dy = \int 4x^4 \, dx$$

Integration is now straightforward, and yields:

$$y = \frac{4x^5}{5}$$

This integration yields the same result as that obtained by integration in the usual way, as described in the previous chapter. The power of this methodology becomes clear when we realize how we can integrate *both* sides of an equation at the same time. ■

Occasionally, we need to rearrange slightly before we can integrate. In such a situation, we ensure that all instances of the controlled variable (including dx) are located on the right-hand side, and all instances of the observed variable (including dy) lie on the left-hand side.

....
Strictly, we must integrate *both* sides.
....

....
The integration on the left-hand side works because the integral of 1 (i.e. $1 \times y^0$) is $1 \times y^1$.
....

■ Worked Example 20.2

A phase diagram depicts values of pressure and temperature at which two or more phases co-exist at equilibrium. Figure 20.1 shows the phase diagram for water. We call the solid lines *phase boundaries*. A modified form of the Clapeyron

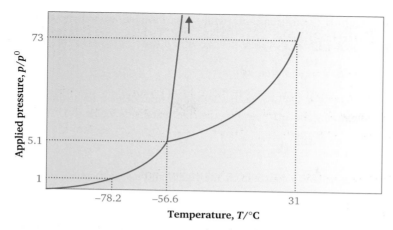

Figure 20.1 The phase diagram of water (not drawn to scale). The **phase boundary** defining the boiling process is depicted in **bold**.

equation quantifies the gradient of the bold phase boundary, which describes the boiling temperature T of liquid water as a function of the applied pressure p:

$$\frac{dp}{dT} = \frac{p\Delta H^{\ominus}_{(vap)}}{RT^2}$$

where R is the gas constant and $\Delta H^{\ominus}_{(vap)}$ is the molar enthalpy of boiling. Rearrange this equation, separate the variables, and then integrate this expression to yield an equation describing the shape of the phase boundary.

Strategy:

(i)   We rearrange this expression, so all instances of the controlled variable (in this case temperature, T) are located on the right-hand side, and all instances of the observed variable (in this case pressure, p) lie on the left-hand side.

(ii) We separate the variables.

(iii) We integrate.

Thus:

(i)   We **rearrange** because the left-hand side is expressed in terms of pressure, but a p term also appears on the right. We therefore cross-multiply with the p term:

$$\frac{1}{p}\frac{dp}{dT} = \frac{\Delta H^{\ominus}_{(vap)}}{RT^2}$$

(ii) We **separate the variables**:

$$\frac{1}{p}\,dp = \frac{\Delta H^{\ominus}_{(vap)}}{RT^2}\,dT$$

(iii) We **integrate**:

$$\int \frac{1}{p}\,dp = \int \frac{\Delta H^{\ominus}_{(vap)}}{RT^2}\,dT$$

We choose to place the $\Delta H^{\ominus}_{(vap)} \div R$ term *outside* the right-hand integral because its value remains constant.

Because R and $\Delta H^{\ominus}_{(vap)}$ are constants, while T is a variable, we will move the constants outside the integral sign:

$$\int \frac{1}{p}\ dp = \frac{\Delta H^{\ominus}_{(vap)}}{R} \int \frac{1}{T^2}\ dT$$

Integrating the reciprocal of p on the left-hand side yields a logarithm (see Chapter 19); and integrating the $1/T^2$ term on the right-hand side yields $-1/T$, so we obtain:

$$\ln p + c = -\frac{\Delta H^{\ominus}_{(vap)}}{R} \times \frac{1}{T} + c'$$

The prime ′ against the constant on the right-hand side is needed because the two constant terms c will differ. In usual practice, we prefer to cite the final result with a single constant, i.e. combine the two c terms, saying:

$$\ln p = -\frac{\Delta H^{\ominus}_{(vap)}}{R} \times \frac{1}{T} + \text{constant} \ \blacksquare$$

Self-test 1

In each case, separate the variables and then integrate the expression:

1.1 $\dfrac{dy}{dx} = zx^3$

1.2 $\dfrac{dy}{dx} = y^2x^2$

1.3 $\dfrac{dc}{de} = \dfrac{c^3}{e}$

1.4 $\dfrac{dh}{dg} = 4h^3 \sin g$

Integration with limits

By the end of this section, you will know:

- We employ the concept of 'limits' if we know the value of a differential at two values of the principal variable.
- Integrating with limits is an alternative way of removing a constant of integration.
- We usually do not write a constant of integration when working with limits.

In the previous chapter, we were able to evaluate a constant of integration if we knew a gradient (i.e. a *differential*) and a single point. At other times, we may know the value of a variable under two separate conditions. We can utilize this situation to circumvent the need to write a constant of integration.

We call this procedure **using limits**. In this method, we first perform an integration procedure, then insert the two (different) values of a variable. In this way, we cause the constant to disappear by subtracting our first answer from the second. The method allows us to evaluate an integral, even if it is indefinite (i.e. it would otherwise require a constant of integration).

We call these two values of 2 and 4, the **limits**.

■ Worked Example 20.3

Evaluate the integral $dy/dx = x^3$ between the values of $x = 4$ and $x = 2$.

Strategy:

(i)   We integrate the expression in the normal way.

(ii) We insert first $x = 4$, then $x = 2$, into the resulting expression.

(iii) We subtract the second result from the first.

Thus:

In words, we would normally rephrase the question to ask us to 'integrate the function x^3 between the limits of 4 and 2'.

(i)   Integrating this differential according to eqn (19.1) yields $\frac{1}{4}x^4 + c$. Because we want limits, we write the expression in square brackets, and write the limits vertically along the right-hand outer side:

$$\int x^3 \, dx = \left[\frac{x^4}{4} + c \right]_2^4$$

(ii) We then insert the values of the limits, the upper value first:

$$\int y = \left[\frac{4^4}{4} + c \right] - \left[\frac{2^4}{4} + c \right]$$

Value of the integral   Value of the integral
when $x = 4$ when $x = 2$

(iii) The two c terms will cancel (which is one reason why we employ this procedure). In fact, we generally do not generally bother to write the constant c because it will cancel.

Subtracting the constant c and inserting the limits transforms the integral:

$$\int y = 64 - 4 = 60$$

The numerical value of the integral is 60. ■

■ **Worked Example 20.4**

Evaluate the integral $\int_0^{\pi/2} \sin 2x \, dx$.

Strategy:

(i)   We integrate the function by rule, employing eqn (19.3).

(ii) We insert first the upper limit then the lower limit.

(iii) We subtract the result with the lower limit from the result with the lower limit.

(iv) Because this function involves trigonometry, we must employ radians.

(i)   Inserting terms into eqn (19.3) yields:

$$\int y \, dx = \left[-\frac{1}{2} \cos 2x \right]_0^{\pi/2}$$

Here, the dx on the left-hand side means we integrate 'with respect to x'.

(ii) We next write the integral twice, and insert the upper limit in the first and the lower limit in the second:

$$\int y \, dx = \left[-\frac{1}{2} \cos \pi + c \right] - \left[-\frac{1}{2} \cos 0 + c \right]$$

(iii) The two values of c will clearly cancel, yielding:

$$\int y\,dx = \left[-\frac{1}{2}\cos\pi\right] - \left[-\frac{1}{2}\cos 0\right].$$

(iv) The cosine of π is the same as the cosine of 180°, so its value is -1. The cosine of 0 is $+1$. Inserting values therefore yields:

$$\int y\,dx = \left[-\frac{1}{2}(\times -1)\right] - \left[-\frac{1}{2}(\times 1)\right]$$

so:

$$\int y\,dx = 1 \ \blacksquare$$

Self-test 2

Evaluate the following definite integrals using limits and eqn (19.1):

2.1 $\displaystyle\int_{1}^{2} x^4 - x^2\,dx$ 2.4 $\displaystyle\int_{0}^{5} x^3 - x^2 + x - 6\,dx$

2.2 $\displaystyle\int_{1}^{3} e^{3x}\,dx$ 2.5 $\displaystyle\int_{9.5}^{10} \frac{2}{x}\,dx$

2.3 $\displaystyle\int_{\pi/3}^{\pi} \cos 2x\,dx$ 2.6 $\displaystyle\int_{0}^{\pi} \frac{\cos 6x}{3}\,dx$

Integration to calculate an area

By the end of this section, you will know:

- Just as we integrate a gradient to obtain the equation of a line, integrating the equation of a line yields an area.
- That integration affords the most accurate method of determining an area.

Notice how we can say the integral is '1' rather than '$1 + c$'.

One of the easier methods of determining the area A beneath a curve is to draw the curve carefully on graph paper, and divide it into thin rectangles. Figure 20.2 shows such a curve: the area beneath the curve and between the two limits A and B is

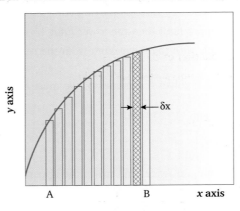

Figure 20.2 The area beneath a curve can be approximated to a series of slender rectangles of height y and width δx.

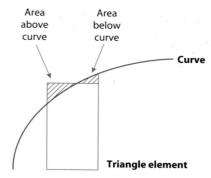

Figure 20.3 To minimize the errors, the area above the curve needs to be the same as the area beneath it. The areas that need to cancel are shaded.

represented by a series of slender rectangles, each with the same width (call it δx), but with differing heights.

A moment's thought suggests that this method is only approximate. The rectangles align with the x axis, but the top of each rectangle overlaps slightly with the curve. With care, the error associated with this method can be minimized: in practice, we ensure the curve cuts neatly across the top of each rectangle, such that the area (by eye) above the curve is the same as the beneath it, see Fig. 20.3. This way, the errors per rectangle then cancel.

We obtain the area of a rectangle by multiplying its height by its width. The width of each rectangle here is δx, and the height is y. We see how the area per rectangle is $y \times \delta x$, and the area between the limits A and B and beneath the curve is the sum of these rectangles: $\Sigma (y \times \delta x)$.

To minimize the errors still further, we ensure that each rectangle is infinitesimally narrow, with a width not of δx but of dx. We say the area is:

$$\text{area} = \sum y \, dx$$

In general practice, we usually write the function of the line in place of the letter y.

■ **Worked Example 20.5**

Calculate the area bounded by the line $y = x^2$ and the x axis and between the vertical lines $x = 3$ and $x = 6$ (the shaded portion in Fig. 20.4).

Methodology:

(i)   We integrate the equation of the curve $y = x^2$.

(ii) As limits, we insert numbers $x = 3$ and $x = 6$.

Thus:

(i)   The function is $y = x^2$, so:

$$\text{area} = \int_3^6 x^2 \, dx$$

(ii) And:

$$\text{area} = \left[\frac{x^3}{3}\right]_3^6 = \left[\frac{6^3}{3} - \frac{3^3}{3}\right] = \frac{216}{3} - \frac{27}{3} = 63$$

The area of the bound portion is 63 units. ■

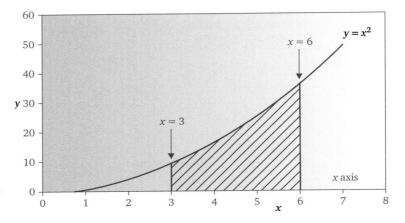

Figure 20.4 We determine the area of the shaded portion by integrating the equation of the curve between the two limits $x = 6$ and $x = 3$.

■ **Worked Example 20.6**

The equation describing the perimeter of a circle (of radius r) is $2\pi r$. What is the area A enclosed within the circle?

We obtain the area of the circle by integrating the equation of the circle, using r and 0 as the limits:

$$\text{area } A = \int_0^r 2\pi r \; dr$$

$$\text{area } A = \left[\pi r^2\right]_0^r$$

$$\text{area } A = \left[\pi \times r^2\right] - \left[\pi \times 0^2\right] = \pi r^2 \quad \blacksquare$$

■ **Worked Example 20.7**

To measure the change in entropy ΔS of a substance as it warms from a temperature T_1 to a higher temperature T_2, we plot a graph of C_p/T (as y) against T (as x), see Fig. 20.5. We then measure the area under the curve. C_p here is the heat capacity. Because the entropy is an area, we can write:

$$\Delta S = \text{area under curve} = \int_{T_1}^{T_2} \frac{C_p}{T} \; dT$$

If the value of C_p for chloroform (**I**) is $70 \, \text{J K}^{-1} \, \text{mol}^{-1}$, calculate the change in entropy when chloroform (**I**) is warmed from 220 to 240 K.

I

We start by rewriting the expression for ΔS:

$$\Delta S = \int_{T_1}^{T_2} C_p \times \frac{1}{T} \; dT$$

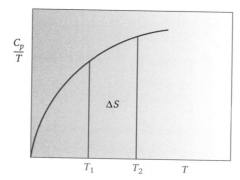

Figure 20.5 The entropy of warming chloroform is obtained as the area under a curve of C_p/T (as y) against T (as x).

In this example, the value of C_p is constant. For this reason, we sometimes place it to the left of the integration sign (we say we place it **outside the integral**). In effect, we place it 'out of the way':

$$\Delta S = C_p \int_{T_1}^{T_2} \frac{1}{T}\, dT$$

We are now in a position to integrate the reciprocal. We know from Chapter 19 that $\int \frac{1}{x}\, dx = \ln x$, so we say:

$$\Delta S = C_p [\ln T]_{T_1}^{T_2}$$

As before, we position the limits vertically along the right-hand outer edge of the square bracket. We next insert the limits:

$$\Delta S = C_p \{[\ln T_2 + c] - [\ln T_1 + c]\}$$

$$\Delta S = C_p [\ln T + c]_{T_1}^{T_2}$$

The two constant terms c will cancel, leaving:

$$\Delta S = C_p \{\ln T_2 - \ln T_1\}$$

Finally, we use the second law of logarithms. We group together the two logarithmic terms:

$$\Delta S = C_p \ln\left(\frac{T_2}{T_1}\right)$$

Finally, to calculate the change in the entropy ΔS of chloroform **(I)** when it is warmed, we insert values into the derived expression to obtain:

$$\Delta S = 70\, \text{J K}^{-1}\, \text{mol}^{-1} \times \ln\left(\frac{240\,\text{K}}{220\,\text{K}}\right)$$

so $\Delta S = 6.1\, \text{J K}^{-1}\, \text{mol}^{-1}$. ∎

> Remember from Chapter 11 how $\log a + \log b = \log(a \times b)$; $\log a - \log b = \log \frac{a}{b}$.

Self-test 3

Determine the areas of the following:

3.1   Between the x axis, the curve $y = x^2 + 4$, and the lines $x = 3$ and $x = 5$.

3.2   Between the x axis, the curve $y = 1/x^2$, and the lines $x = 6$ and $x = 2$.

Additional problems

....
This definition derives
ultimately from the
so-called **Maxwell
relation**, which relates
the differential of the
Gibbs function to volume.
....

20.1 The Gibbs function G changes when the pressure of an ideal gas is changed, according to the relation:

$$\Delta G = \int_{p_1}^{p_2} \frac{1}{p}\, dp$$

Perform this integration to obtain an expression for ΔG.

20.2 Consider a fixed amount of gas in a cylinder. The work w performed when the volume V of a gas is altered is given by the expression:

$$w = \int_{V_{(initial)}}^{V_{(final)}} dV$$

Perform this integration to obtain an expression for w.

20.3 Current I is defined as the rate of change of charge Q:

$$I\frac{dQ}{dt}$$

where t is time. Separate the variables and integrate this expression to obtain an equation with Q as the subject.

20.4 When the reaction of a chemical A proceeds via second-order kinetics, the rate of change of $[A]_t$ follows the equation:

$$\frac{d[A]}{dt} = -k[A]^2$$

Rearrange, separate the variables, then integrate this expression to obtain the integrated second-order rate equation.

20.5 Acceleration a is defined as the rate of change of velocity v:

$$a = \frac{dv}{dt}$$

Integrate this expression using these limits. To obtain the constant of integration, the velocity at time $t = 0$ is u and the velocity at time $t = t$ is v.

20.6 Evaluate the following definite integral:

$$\int_2^4 \sqrt[3]{5x}\, dx$$

20.7 The speed at which electrons move through a semiconductor relates to its mobility μ. For many semiconductors, a gradient of a graph of temperature T against mobility μ has the following form:

$$\frac{d\mu}{dT} = k\sqrt{T}$$

where k represents a collection of constants. Separate the variables and hence integrate this expression to obtain a relationship in which μ is the subject.

....
For convenience,
position is defined
here as distance l from
the start position.
....

20.8 We define an object's velocity v as the rate at which its position changes:

$$v = \frac{dl}{dt}$$

Separate the variables and hence integrate this expression to obtain a relationship involving v, l, and t.

20.9 When the reaction of a chemical A proceeds via first-order kinetics, the rate of change of $[A]_t$ follows the equation:

$$\frac{d[A]}{dt} = -k[A]$$

Rearrange, separate the variables, then integrate this expression to obtain the integrated first-order rate equation.

20.10 Evaluate the following definite integral:

$$\int_{\pi/12}^{\pi/6} \sin 4x \; dx$$

Statistics I

Averages and simple data analysis

The concept of a data spread

By the end of this section, you will:

- Know that real experimental data are not exact.
- Understand that data will follow a Gaussian ('normal') distribution.
- Know the difference between 'accuracy' and 'precision'.

When we measure things in the laboratory, the reading we obtain will never be exact. Some pieces of equipment give better results than others—a modern top-pan balance is likely to give a trustworthy result; and undergraduate experiments involving calorimetry are notoriously poor at yielding the results similar to those in authoritative books of data.

Chemistry textbooks almost always contain tables of data. Chemists generally call such data 'literature values'. It is easy to think of such values as being '*the* value', i.e. entirely correct, which implies that we would obtain exactly the same result if only we were more careful. This assumption is not always safe. The reasons for differences between our own values and those in the literature could include:

- The data in such books were *computed* rather than measured experimentally.
- The equipment used to obtain the data was more sophisticated, or completely different.
- The scientists obtaining the data took additional precautions.
- The data in the book are in fact wrong.

While it is always wise to compare our experimental data with those in the literature, it is also wise to critique the literature data, asking how they were obtained. For example, when we look closely at the data in such books, we generally see provisos have been made for us. For example, the enthalpy of solution for potassium nitrate is cited as $34.89 \pm 0.02\,\text{kJ}\,\text{mol}^{-1}$. The additional term '$\pm 0.02$' tells us the legitimate value is in fact a range: it can extend from '$34.89 - 0.02$' $\text{kJ}\,\text{mol}^{-1}$ through to '$34.89 + 0.02$' $\text{kJ}\,\text{mol}^{-1}$. Only if our value is outside this range should we question our result.

In this chapter, we explore the simpler ways of treating experimental data. Firstly, we see if we can trust them, i.e. we make a qualitative assessment; and, secondly, we make a quantitative assessment, i.e. we decide the extent to which we wish to accept them.

Competent chemists never perform a single analysis, but seek to *replicate* the data: that is, they perform the measurement several times and obtain a *series* of results, which we call a data set. Once this series is available, the chemist will then assess the data statistically. Only in this way will their reliability become apparent.

The central idea behind the use of statistics lies in the way a series of data can differ. For example, even good-quality data will show a spread of results: we should expect it. Typically, the experimental data will show a spread described by a Gaussian distribution, like that in Fig. 21.1.

The concept of 'data spread' lies behind the Gaussian distribution curve. The *x* axis coves the full range of data values, while the notation on the *y* axis, $n(x)$, is the frequency with which these values occur. The centre of the peak corresponds to the

---

Reminder: The word *data* is a plural; the singular noun of data is *datum* and never '*datas*'

In words, we say the symbol $\pm$ means 'plus or minus'.

The words **quantitative** and **qualitative** are explored in depth in Chapter 7.

A Gaussian distribution is often called a **normal distribution**, where the qualifying term 'normal' warns us that we should *expect* our data to follow a similar pattern to that in Fig. 21.1. Data which do not follow such a distribution may be suspect.

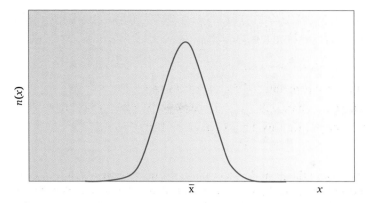

Figure 21.1 A schematic representation of a Gaussian distribution of data, being a plot of number of data having a value of x (as y) against x. The peak occurs at a value of $\bar{x}$.

average value of the data set (and the full meaning of 'average' will be explored in subsequent sections). The 'bell' shape of the figure shows how most data will have values at (or near) this average value. In other words, the majority of the results will be similar. But some of the data have values which are considerably different. We give the name **outlier** to the points that deviate. The shape of the curve tells us that outliers that deviate by a large extent are rare; data that deviate by a smaller proportion are common.

Because our data will normally follow a Gaussian distribution, and because we want our final results to be reasonable, we may wish to follow the example of many analysts and 'pre-select' data before we start any statistical analyses. We look at methods for pre-selection when we introduce the standard deviation, below.

We now look further at the Gaussian distribution curve to see what it can tell the analyst. Firstly, the shape of the distribution curve in Fig. 21.1 can be shallow and wide or narrow and sharp. We describe this aspect as the **precision** of the data. Secondly, the position of the central peak may coincide with the correct value of the variable we wish to quantify, or may be completely different: for example, are there significant factors we failed to accommodate during our preparation? The difference between the position of the peak and the true value is termed the **accuracy**.

It is very common to see the terms **precision** and **accuracy** used interchangeably. Such a practice is incorrect.

We will employ a different analogy to explore the similarities and differences between 'precision' and 'accuracy'. Think of a dartboard: it will be circular, and marked with concentric circles. The exact centre of the board corresponds to the true value of the analysis. The position of each dart (as marked by a black dot) indicates a value obtained by an analyst. Figure 21.2 looks at the four common outcomes.

The four 'snapshots' in Fig. 21.2 clearly depict different situations. Analysts want measurements of high accuracy: that is, with the data centred round the correct value. The dartboards in traces (a) and (c) both depict analyses of high accuracy. In trace (a), the points are widely scattered, so they are imprecise. The precision is low. Conversely, in trace (c), the points are clustered more tightly around the centre, implying a higher precision. We will quantify the precision in later sections of this chapter.

Traces (b) and (d) depict analyses of poor accuracy. Even in trace (d) where all the points are positioned closely together, they are positioned away from the centre of the board which represents the correct value. Trace (b) is even worse: the accuracy is low and the precision is also poor. The traces in Fig. 21.2(a)–(d) are extremes. In reality, the precision and accuracy may be intermediate.

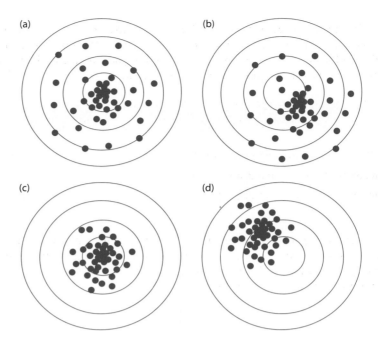

Figure 21.2 'Hits on a dartboard': (a) high accuracy and low precision; (b) low accuracy and low precision; (c) high accuracy and high precision; and (d) low accuracy and high precision.

There are usually many causes of poor accuracy, such as human error (rushing or poor technique), but also the innate accuracy of the apparatus we use. The errors are **random** or **indeterminate**. The data set will not necessarily follow a Gaussian distribution if the data are truly random.

The cause of poor precision is more likely to be the equipment we employ. For example, the dial on a flame photometer is consistently too high, a pipette was incorrectly calibrated, or our equation is not balanced correctly. We call such an error a **systematic** or **determinate error**, because (in principle) we can locate it and/or compensate for it.

We now look at quantitative methods of analysing our data set.

Averages

By the end of this chapter, you will know:

- The word 'average' is a statistical construct.
- There are three common types of average: the mean, the mode, and the median.

Because an analyst will make multiple measurements, and obtain a series of results, the simplest way of obtaining 'an answer' is to take the average of the data (or, strictly, of a *series* of data). But the word 'average' is ambiguous, so we will deliberately avoid it on its own in this chapter. Where it does appear, its use will always be qualified.

The ambiguity arises because 'average' can mean one of three distinct types of calculation: the **mean**, the **mode**, and the **median**. In everyday language, 'average' generally implies the mean, which explains why, if we type 'average' as the function in an Excel™ spreadsheet, the program computes a mean.

The mean

By the end of this section, you will know:

- The mean is also called the *arithmetic* mean.
- The mean is a weighted average of a set of data.

The word 'average' generally means the mathematical function known as a **mean**. Such a 'mean' is a statistical way of looking at a series of data, and allows us to represent a *series* of data with a single value.

The mean is calculated using eqn (21.1):

$$\bar{x} = \frac{\sum_{i=1}^{N} x_i}{N} \tag{21.1}$$

In words, we add up all the data, and divide by the number of data in the set. Notice the small bar positioned over the x, as $\bar{x}$. This bar tells us at a glance that we have a mean. The value we compute will always lie between the extremes of the data we employed when computing the mean, so $\bar{x}$ will always lie between the largest and smallest data points. We have miscalculated $\bar{x}$ if its value does not lie within this range.

■ **Worked Example 21.1**

An analytical chemist wishes to determine the accuracy of a $1.0 \, cm^3$ pipette. The method chosen is to measure out $1.0 \, cm^3$ of water 10 times, and weigh the amount of water in each case. In grams, the mass of water is 1.003, 0.983, 1.022, 1.031, 1.011, 1.001, 0.939, 0.993, 0.992, and 0.988 g.

If the density of water is $0.998 \, cm^3 \, g^{-1}$, what is the mean mass of water $\bar{m}$ and hence what is the mean volume $\bar{V}$?

Inserting data into eqn (21.1) yields:

$$\bar{m} = \frac{(1.003 + 0.983 + 1.022 + 1.031 + 1.011 + 1.001 + 0.939 + 0.993 + 0.992 + 0.988)g}{10}$$

so:

$$\bar{m} = \frac{9.963 \, g}{10} = 0.9963 \, g$$

To three significant figures, the volume of the pipette is therefore:

$$\bar{V} = \bar{m} \times \text{density, so } \bar{V} = 0.9963 \, g \times 0.998 \, cm^3 \, g^{-1} = 0.994 \, cm^3$$

In other words, the pipette is 0.6% smaller than it claims to be. ■

The mean is also called the **arithmetic mean**, because we calculate it as a *sum* of the component data. A different mean is the **geometric mean**. Chemists only rarely employ the *geometric* mean, which is defined in terms of a Pi product Π (rather than the Sigma sum in eqn (21.1)), so we will not consider it further.

Note that it is common and acceptable to cite a mean to more **significant figures** than the original data within the set. In general, the mean is cited to $(n + 1)$ s.f.s if the data are cited to n s.f.s.

The mode

By the end of this section, you will know:

- The mode is the value which occurs most frequently.
- Statistically, the mode is rarely as useful as the mean.

For the analyst, the mode is not generally as useful as the mean, and is not employed so often. As before, the analyst measures a variable several times to obtain a series. The mode is then the value which occurs most frequently. In other words, if we draw a bar graph, the mode corresponds to the highest bar.

■ **Worked Example 21.2**

An industrial effluent is found to contain the banned insecticide dichlorodiphenyl trichloroethane, DDT (**I**). The amount of [DDT] (in parts per million, ppm) was found to be 21.2, 21.3, 20.9, 21.4, 21.3, 21.7, 20.6, 20.3, 22.3, 21.4, 21.5, 21.3, 21.0, 21.5, 21.3, 21.2, and 20.6.

Display the data as a bar graph, and hence determined the mode. ■

The highest peak in the bar chart in Fig. 21.3 is that for [DDT] = 21.3 ppm, so the mode is 21.3 ppm. As expected, the mode lies at the centre of the spread of data. Note

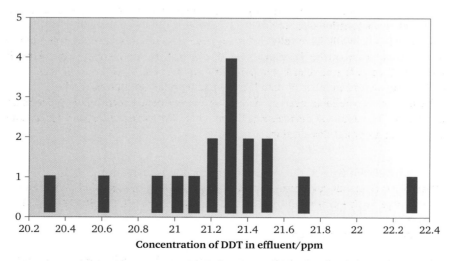

Figure 21.3 A bar graph of the concentration of DDT (**I**), an industrial effluent (as x), against number of times this concentration was obtained in an analysis (as y).

how the shape of the graph in Fig. 21.3 broadly follows the Gaussian distribution in Fig. 21.1.

I

Self-test 2

Determine the mode of each series of data:

2.1   mass (in mg): 9.21, 9.19, 9.20, 9.18. 9.21, 9.17, 9.20, 9.18, 9.22, 9.24, 9.20, and 9.21

2.2   concentration (in ppm): 212, 213, 212, 215, 219, 221, 211, 227, 220, 215, 212, 215, 214, and 213

2.3   concentration in $\mu g\,dm^{-3}$: 32, 33, 32, 34, 35, 33, 34, 33, 30, 29, 31, 32, 33, and 30

The median

By the end of this section, you will know:

- The median of a range of data lies in the middle of the spread of values.
- We can only properly take the median of an odd number of data.
- If we want the mode of an even number of data, we take the mean of the *two* central data.

The median is another form of average. It is rarely used by the analyst because its value is often of questionable worth.

Any series of data will have values with a spread of values. The median is defined as the middle value. Accordingly, if there are five data, then the median is the third; if there are 99 data, the median is the 49th, and so on. Clearly, we must be strict and employ only an odd number of data if a median is to be determined.

The median is a reliable average if there are many data which follow a Gaussian distribution. Its usefulness decreases as the number of data decreases, and as the data deviate from a normal distribution.

■ **Worked Example 21.3**

Several students sit an exam in analytical chemistry. In order of increasing percentage, their exam marks are 12, 22, 30, 40, 45, 50, 53, 55, 56, 58, 60, 70, and 90%.

What is the median mark?

There are 13 marks, so the median corresponds to the middle one, the seventh. Accordingly, the median mark is 53%.

12, 22, 30, 40, 45, 50   **53**   55, 56, 58, 60, 70, 90

six data *before* the median   the median   same number of data
after the median

Roughly half of our multiple analyses will yield a data set comprising an even number. To obtain the mode of such a data set, we again arrange the data in ascending order, and then choose the *two* central points. The median is then the mean of these two.

Self-test 3

Determine the median of each data set. [Hint: always place the numbers in increasing order before starting.]

3.1 The length of a sample, in cm: 1.90, 1.93, 1.90, 1.98, 1.99, 1.93, 1.93, 1.90, and 1.93.

3.2 The energy of an N=N bond, e.g. in azine dyes, in $kJ\,mol^{-1}$: 473, 474, 472, 476, 475, and 475.

The standard deviation

By the end of this section, you will:

- Appreciate how the sample standard deviation relates to the relative precision of a result.
- Know how to calculate the sample standard deviation *s*.
- Know that a high value of *s* implies poor precision.

Indeterminate (random) errors can normally be analysed by simple statistical methods. We now understand that chemical analysts measure a variable several times, and that the results will vary. Having looked at averages, i.e. at ways of abbreviating a series to form a single number, we now look at the validity of that single number, asking 'how accurate is the mean: how wide is the spread of results within the data set?'

Most of the simpler statistical methods assume the data set follows a Gaussian distribution profile. The first statistical tool is the **sample standard deviation *s***. By the very nature of the sample standard deviation, a large value of *s* means a wide spread of data and a smaller value of *s* means a narrow spread. In other words, large values of *s* imply imprecision, and a small value of *s* means a more precise data set. To reiterate, the precision tells us nothing about the accuracy: an answer may be precise but wrong.

In terms of a formal definition, the sample standard deviation *s* is:

....
The sample standard deviation *s* describes the spread of data around the mean datum within a data set.
....

$$s = \frac{\sqrt{\sum_{i=1}^{N}(x - \bar{x})^2}}{N - 1}$$

(21.2)

where N, the number of data within the set, is typically less than 10, and $\bar{x}$ is the mean.

This definition looks rather complicated, but mathematical rearrangement allows us to rewrite it in a more 'user-friendly' way:

$$s = \sqrt{\frac{\sum\limits_{i=1}^{N} x_i^2 - \left(\sum\limits_{i=1}^{N} x_i\right)^2 / N}{N-1}} \tag{21.3}$$

This version is considerably easier to use, because we calculate two sums, and then take the root of their difference.

Also notice how s has the same units as the original variable, so if our data set relates to concentrations in parts per billion (ppb), then we also express s in ppb.

*This definition involves taking the square root of an arithmetic mean of a series of square terms. This three-fold procedure is a common feature in statistics, so we often abbreviate it saying, **root mean square**.*

■ **Worked Example 21.4**

An electrochemist repeatedly measured the capacitive charging current $i_{(cap)}$ (in µA) as: 19.4, 20.6, 18.7, 19.2, 21.6, 18.9, and 19.9.

Calculate the sample standard deviation s of this data set

Strategy:

We calculate each term separately:

(i) $\sum\limits_{i=1}^{N} x_i^2$.

(ii) $\left(\sum\limits_{i=1}^{N} x_i\right)^2$.

(iii) Lastly, we assemble the terms within eqn (21.3).

Thus:

(i) To calculate $\sum\limits_{i=1}^{N} x_i^2$:

$$\sum\limits_{i=1}^{N} x_i^2 = (19.4)^2 + (20.6)^2 + (18.7)^2 + (19.2)^2 + (21.6)^2 + (18.9)^2 + (19.9)^2$$

$$= 376.36 + 424.36 + 349.69 + 368.64 + 466.56 + 357.21 + 396.01$$

$$= 2738.83$$

(ii) To calculate $\left(\sum\limits_{i=1}^{N} x_i\right)^2$:

$$\left(\sum\limits_{i=1}^{N} x_i\right)^2 = (19.4 + 20.6 + 18.7 + 19.2 + 21.6 + 18.9 + 19.9)^2$$

$$= (138.3)^2$$

$$= 19\,126.89$$

(iii) There are seven items of data. Inserting these numbers within eqn (21.3) yields:

$$s = \sqrt{\frac{2738.83 - (19\,126.89)/7}{7-1}}$$

$$s = \sqrt{\frac{2738.83 - 2732.42}{6}}$$

$$s = 1.1 \text{ µA}$$

Notice how the standard deviation has the same units (here µA) as the data it refers to.

The magnitude of the sample standard deviation here is not large—about 5% of the mean (in this example, $\bar{x} = 19.76$ µA). ■

We are generally concerned when an experiment yields a value of s of 20% or more; but occasionally, we actually want a large value of s. For example, the standard deviation for an examination ought to have a high value of s to demonstrate a clear separation between the students on this course.

When a data set comprises a larger number of data points (typically greater than 10), we employ a slightly different version of s, known as the **population standard deviation** σ. Note how the two standard deviations have different symbols.

The two expressions in eqns (21.3) and (21.4) describe essentially the same quality, i.e. the variations within a data set.

$$\sigma = \sqrt{\frac{\sum\limits_{i=1}^{N} x_i^2 - \left(\sum\limits_{i=1}^{N} x_i\right)^2 / N}{N}} \qquad (21.4)$$

The only mathematical difference between eqns (21.3) and (21.4) lies in the denominator within the root: because there are more data within the set, we divide by the total number N. We say the expression in eqn (21.4) has an **extra degree of freedom**.

We only rarely replicate a measurement more than 10 times, so the sample standard deviation s is encountered and used far more frequently than the population standard deviation σ.

Aside

The calculation in Worked example 21.4 is long-winded, time consuming, and prone to errors and slips. Luckily, we can speed up the process by using either a calculator or a computer spreadsheet.

Using a pocket calculator

Some 'scientific' calculators will calculate the standard deviations automatically. It's always wise to read the calculator manual before use, because every calculator is different. The necessary keys will generally have the labels $\boxed{\sigma_n}$ and $\boxed{\sigma_{n-1}}$. Some calculators use a more advanced notation, with two sets of keys: labels $\boxed{x\sigma_n}$ and $\boxed{x\sigma_{n-1}}$, and $\boxed{y\sigma_n}$ and $\boxed{y\sigma_{n-1}}$.

Using a computer spreadsheet

Ensure there is no gap after the = sign here. The letters are not case sensitive.

(i)   Open Excel™.

(ii) Enter the data in a single column.

(iii) In a cell below the column, type '=STDEVA()'.

(iv) Press RETURN twice, i.e. cancelling the 'error' message.

(v)   Using the mouse, highlight all the cells to be considered.

(vi) Press RETURN once more.

The cell should display the standard deviation. (It may be necessary to change manually the number of significant figures.)

Self-test 4

Determine the standard deviations of the two data series in Self-test 3.

The Q-test

By the end of this section, you will:

- Know that outlier points are statistically rare.
- Know that including outlier points in a calculation can ruin the result.
- Know how to use a simple Q-test to identify outlier points.

It is now clear how the data within a set have a range of values. The majority have a value close to the mean, a few could have values which are extreme and seriously distort a mean or standard deviation calculation. Such data must be identified and may be omitted before they distort the final result. It must be stressed, though: *great care is needed when rejecting data because we effectively introduce a bias each time we reject a datum.*

We need to have confidence in our data: are they reliable? If not, then we are foolish if we continue using them. Unfortunately, there is no guaranteed method for rejecting or retaining data, but the simplest method, which assesses the reliability of a single datum at a time, is the so-called *Q-test*.

In the *Q-test* method, we first calculate the quotient Q of a single experimental point, according to the equation:

$$Q_{(exp)} = \frac{x - x_{(next)}}{x_{(maximum)} - x_{(minimum)}} \qquad (21.5)$$

where the value of x without a subscript in the numerator relates to the point under scrutiny. The other three terms relate to other members of the data set, which have already been arranged in order of increasing value: $x_{(next)}$ is the value of the datum next in the series, and (in the denominator), $x_{(minimum)}$ and $x_{(maximum)}$ are respectively the data of the two extreme ends of the set.

Having determined a value of $Q_{(exp)}$, we compare its value with calculated data in a table. Table 21.1 contains such a compendium, as explained in Worked example 21.5.

■ **Worked Example 21.5**

A chemist determines the absorbance of the dye Indigo Carmine (**II**) solutions in order to determine its concentration. The absorbances A within the data set are 0.104, 0.113, 0.114, 1.114, 0.115, and 0.117. We suspect the first point is anomalous.

Using a *Q-test*, with what confidence can the first point be rejected?

II

Strategy:

(i)   Calculate $Q_{(exp)}$ using eqn (21.5).

(ii) Compare the value of $Q_{(exp)}$ with values in Table 21.1.

Thus:

(i)   The suspect datum is 0.104; the neighbouring point $x_{(next)}$ has a value of 0.113; the maximum point $x_{(maximum)}$ is 0.117; and the minimum point $x_{(minimum)}$ is 0.104 again. Inserting values into eqn (21.5) yields:

$$Q_{(exp)} = \frac{0.104 - 0.113}{0.117 - 0.104} = -\frac{0.009}{0.013}$$

We ignore the sign. Therefore, $Q_{(exp)} = 0.692$.

(ii) We now look at Table 21.1.

Table 21.1 Q-test table, indicating confidences for replicate measurements, allowing for the confident assessment of outlying data.

No. of replicate measurements	Reject with 90% confidence	Reject with 95% confidence	Reject with 99% confidence
3	0.941	0.970	0.994
4	0.765	0.829	0.926
5	0.642	0.710	0.821
6	0.560	0.625	0.740
7	0.507	0.568	0.680
8	0.468	0.526	0.634
9	0.437	0.493	0.598
10	0.412	0.466	0.568

Our data set comprises six values, so we will concentrate our attention to the row starting '6'. If $Q_{(exp)}$ had a value of 0.56, we could have rejected it with 90% certainty; in other words, we are uncertain about the value 10% of the time. Conversely, if the value of $Q_{(exp)}$ was 0.625, we would have 95% certainty the value was erroneous, so we would have halved the uncertainty. And if $Q_{(exp)}$ had a value of 0.740, we would have 99% confidence the datum was an outlier which should be rejected. The uncertainty is only 1%. The higher the value of $Q_{(exp)}$, the safer we are when rejecting a datum.

To return to Worked example 21.5, the value of $Q_{(exp)}$ was 0.692. We can reject this datum with a confidence of more than 95%, but with a confidence of less than 99%. In other words, the point is almost certainly unreliable, so we can safely reject it before any statistical manipulation of the data. ∎

So, in summary, before we manipulate the data, calculating mode, mean, and median, we need to ascertain whether the data are reliable, and omit those that are likely to be suspect.

More comprehensive tables are available if we want to know the confidence to other degrees of certainty.

Self-test 5

Each of the following data sets contains a suspect point. Identify the suspect point, and then use a simple Q-test to ascertain the confidence with which we could reject the suspect point.

5.1 Using a Karl–Fischer titration apparatus, the amount of water in a sample of ethanol is: 0.65, 0.68, 0.71, 0.72, and 0.92%.

5.2 Five samples of an ore were analysed for their lead content. The content, in ppm, is: 214, 217, 219, and 226.

5.3 A chemist titrates the acid in the gut of a mouse, and determines its concentration (in $mmol\,dm^{-3}$) as: 32.2, 34.3, 35.2, 35.2, 35.4, and 35.7.

Additional problems

Thirteen students each perform an acid–base titration. Their titres of alkali solution are: 17.0, 18.3, 18.4, 18.5, 18.8, 18.8, 18.9, 18.9, 18.9, 19.2, 19.3, 19.5, and $20.0\,cm^3$.

21.1 Perform a simple Q-test analysis of the data.

21.2 Using those data that we are confident are reliable, calculate the mean titre.

21.3 Using those data that we are confident are reliable, calculate the standard deviation associated with the titre.

21.4 Using those data that we are confident are reliable, calculate the median titre.

21.5 Using those data that we are confident are reliable, calculate the mode titre.

A thermodynamicist determined the enthalpy change associated with the reaction:

$$2H_2S_{(g)} + SO_{2(g)} \rightarrow 2H_2O_{(g)} + 3S_{(s)}$$

and found the enthalpy change associated with reaction was -230.0, -230.2, -230.2, -230.3, -230.6, -230.6, -230.6, -230.8, -230.9, and $-232.7\,kJ\,mol^{-1}$.

21.6 Perform a simple Q-test analysis of these data.

21.7 Using those data that we are confident are reliable, calculate the mean energy.

21.8 Using those data that we are confident are reliable, calculate the standard deviation associated with the energy measurement.

21.9 Using those data that we are confident are reliable, calculate the median energy.

21.10 Using those data that we are confident are reliable, calculate the mode energy.

Statistics II

Treatment and assessment of errors

The readings and errors, and the concept of signal-to-noise

By the end of this section, you will learn:

- That all real experimental data have an associated error.
- Because of this error, all readings actually represent a *range* of values.
- The meaning of signal-to-noise ratio.
- All experimental apparatus has a recommended range of values.
- Analysts should ensure the experimental readings lie within the recommended range of the apparatus.

Any reading ever made by a chemist has an associated error. The error may be so small that we can effectively ignore it, but the error may be so large that we must seriously question the meaningfulness of the reading.

We usually cite a reading in the form given in eqn (22.1):

$$\text{reading} = \text{value} \pm \text{error} \tag{22.1}$$

....

Reminder: In words, we say $\pm$ means 'plus or minus'.

....

The symbol $\pm$ means the reading actually possesses a range of values from 'value − error' through to 'value + error'.

■ **Worked Example 22.1**

The enthalpy of the reaction:

$$Fe^{3+}_{(aq)} + SCN^{-}_{(aq)} \rightarrow [FeSCN]^{2+}_{(aq)}$$

is $\Delta H_r^{\ominus} = -35.5 \pm 0.7 \, \text{kJ mol}^{-1}$. What are the range of values for $\Delta H_r^{\ominus}$?

The minimum value of $\Delta H_r^{\ominus}$ is $-35.5 - 0.7 \, \text{kJ mol}^{-1} = -36.2 \, \text{kJ mol}^{-1}$
The maximum value of $\Delta H_r^{\ominus}$ is $-35.5 + 0.7 \, \text{kJ mol}^{-1} = -34.8 \, \text{kJ mol}^{-1}$. ■

We sometimes refer to the relative magnitudes of the reading and the error in terms of the **signal-to-noise ratio**, where the reading is the signal. The ratio is occasionally symbolized as SN or S/N. The ratio is determined simply as the quotient in eqn (22.2):

$$\text{signal-to-noise ratio} = \frac{\text{reading}}{\text{error}} \tag{22.2}$$

....

We usually indicate a **ratio** in the form 'number:1'. The colon tells us the number is a ratio; the last number is always 1.

....

The signal-to-noise ratio is often cited as a ratio, i.e. as 'signal to noise:1'.

There are no absolute rules, but most scientists would say a signal-to-noise ratio of 3:1 or less means the reading is too unreliable to use.

■ **Worked Example 22.2**

A solution of the intensely blood-red-coloured complex $[FeSCN]^{2+}_{(aq)} = 0.05$ mol dm^{-3} is prepared (cf. Worked example 22.1), and its optical absorbance determined to be 3.2 ± 1.2; see Fig. 22.1.

What is the signal-to-noise ratio?

Inserting numbers into eqn (22.2):

$$\text{signal-to-noise ratio} = \frac{3.2}{1.2} = 2.67$$

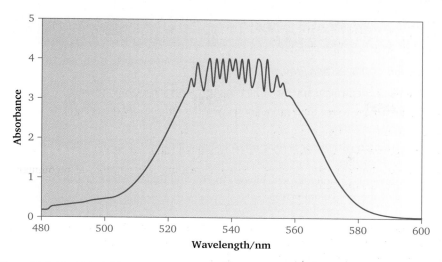

Figure 22.1 The UV–visible spectrum of $[FeSCN]^{2+}_{(aq)} = 0.05\,mol\,dm^{-3}$. The jagged appearance at the top, near the spectrum centre is a manifestation of the large signal-to-noise ratio, and is caused by the detector 'not coping' with the extremely large absorbance.

So the signal-to-noise ratio is 2.67:1. In other words, the error is so large that we ought to obtain the reading in some other way. ■

Equation (22.2) suggests the best way to enhance the signal-to-noise ratio is to ensure the reading is huge, even if the error is large. Occasionally, however, sensitive analytical equipment cannot cope with large readings, meaning that we actually worsen the signal-to-noise ratio. We discern how any piece of equipment has an optimum range: below a certain threshold the reading is too small, and above a different threshold the error is too large.

An experienced analyst will always look up the extremities of this range before making a measurement. If the reading lies outside this recommended range, the analyst should (i) employ a different method, or (ii) modify the analyte (dilute it, pre-concentrate it, etc.) to change the reading, so it enters the recommended range. For example, in Worked example 22.2, we should modify the dye solution until its absorbance is considerably smaller by diluting it, since the Beer–Lambert law suggests that absorbance is directly proportional to concentration.

In optical spectroscopy, the best way to minimize the signal-to-noise ratio is to ensure the absorbance reading has a magnitude in the range 0.75 to 1.0.

Self-test 1

Calculate the signal-to-noise ratio, and decide whether the reading is usable:

1.1 $\Delta H_r^{\ominus} = 4.6\,kJ\,mol^{-1}$ error $= 1.9\,kJ\,mol^{-1}$

1.2 $\lambda_{(max)} = 556\,nm$ error $= 20\,nm$

1.3   absorbance $= 2.5$ error $= 0.02$

1.4   transmittance $= 89\%$ error $= 3\%$

1.5 $emf = 1.104\,V$ error $= 12\,mV$ (note the standard factor)

1.6   time $= 34\,s$ error $= 0.5\,s$

The magnitude of an error

By the end of this section, you will learn:

- The error represents a spread of legitimate values obtained experimentally.
- If the data have a Gaussian spread of values, the error is half the separation between the extreme values.
- When a single parameter is measured, the minimum error is the quotient of innate error and reading.
- When several parameters are measured, the minimum error is the 'root mean square' of the individual errors.

The simplest way to find the magnitude of an error is to read the manual written for a particular instrument. But sometimes the manual is either unhelpful or merely not available. We need to determine the error for ourselves.

■ Worked Example 22.3

An electrochemist measures the *emf* of a cell using a digital voltmeter (DVM). The reading of *emf* fluctuates continually between 110 mV and 130 mV, as shown by a graph of *emf* (as *y*) against time (as *x*); see Fig. 22.2. What is the error?

In this simple example, the *emf* fluctuates with time. The reason for the fluctuation could be merely the random behaviour of the electronics within the DVM, or the movement of air over the cell in the laboratory, stirring the solution.

While the *emf* varies between 110 and 130 mV, the median value is 120 mV, so the *emf* fluctuates between '120 − 10 mV' and '120 + 10 mV'. The error in the *emf* is therefore 10 mV. In this example, we could have written $emf = 120 \pm 10$ mV.

In this example, the spread of data is 20 mV, and the error is 10 mV. In fact, whenever the data follow a Gaussian distribution of values, the error will be half the

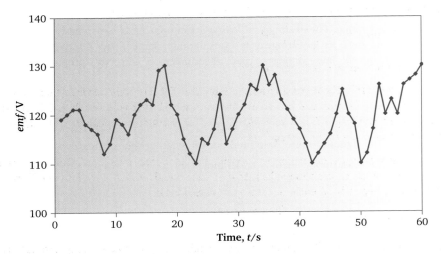

Figure 22.2 The *emf* on a voltmeter can fluctuate with time in a more or less random manner.

separation between the extreme values, provided the extremes are statistically acceptable (for details, see Chapter 21). ∎

Occasionally, we don't have a fluctuation. Even when the reading appears unchanging with time, it has an **innate** (or 'intrinsic') error.

> ■ **Worked Example 22.4**
>
> What is the innate error when measuring a length with a pocket ruler on which the separation between the marks is 1 mm?

Because the marks on the ruler are placed at 1 mm intervals, we cite a length to the nearest mm. Therefore, each time we cite a length l, we have made a decision, choosing which of the marks on the ruler is closest to the real length, l. If this length is closest to the 11 mm mark, we say $l = 11$ mm, but if the length is minutely longer and now closer to the 12 mm mark, we say $l = 12$ mm.

The shortest distance we can choose with confidence is 1 mm, so the innate error is half this distance, i.e. 0.5 mm. ∎

This outcome is general: when measuring a variable with a pre-calibrated instrument, the innate error is half the separation between the gradations on the scale.

But we need now to decide the magnitude of an error. Equation (22.3) shows how we calculate the minimum error associated with a reading:

$$\text{minimum error} = \left(\frac{\text{innate error}}{\text{reading}} \right) \qquad (22.3)$$

Clearly, we wish to minimize the error, so we aim for a small innate error and employ an instrument with a finely divided scale, and we aim for a large reading.

> ■ **Worked Example 22.5**
>
> A chemist is diluting a solution 10-fold. To this end, 1 cm³ of solution is discharged from a burette and made up to 10 cm³. What is the error if the burette divisions are 0.1 cm³?

The gradations are 0.1 cm³ apart, so a competent chemist will work with an error of 0.05 cm³. To obtain the error, we insert figures into eqn (22.3):

$$\text{minimum error} = \left(\frac{0.05}{1.0} \right) \text{cm}^3 = 0.05 \text{ cm}^3$$

$$\text{minimum error} = 0.05 \text{ cm}^3, \text{ or } 5\%$$

The error will probably be considerably greater than 5% if the chemist is inexperienced.

The best way to minimize this error is to work with larger volumes than 1.0 cm³, so rather than running 1.0 cm³ to make 10 cm³ of solution, the chemist should run 10.0 cm³ and make up to 100 cm³. ∎

In many instances, chemists need to determine several parameters, and then perform a calculation. Each measurement has its own associated errors; and the overall minimum error depends on each, according to eqn (22.4):

$$(\text{minimum error})^2 = \sum_{i=1}^{n} \left(\frac{\text{innate error}}{\text{reading}} \right)^2 \qquad (22.4)$$

This sum is conceptually similar to the calculation performed using eqn (22.3), above.

■ **Worked Example 22.6**

In order to determine the solubility product of silver chloride, a chemist titrates $AgNO_3$ (0.1 mol dm^{-3}) against aqueous KCl (0.1 mol dm^{-3}), and measures the change in *emf* after the addition of each aliquot. Each aliquot has a volume of 1.0 cm^3 and the burette scale has divisions of 0.1 cm^3; the *emf* at the end point is 0.255 V and the minimum division on the digital voltmeter (DVM) is 1 mV; and the mass of KCl was 0.745 g on a top-pan balance whose LCD cites the mass to ±0.1 g.

What is the error in the measurement?

We first rewrite eqn (22.4):

$$(\text{minimum error})^2 = \left(\frac{\text{burette error}}{\text{burette volume}}\right)^2 + \left(\frac{\text{DVM error}}{\text{DVM reading}}\right)^2 + \left(\frac{\text{mass error}}{\text{overall mass}}\right)^2$$

The overall minimum error is therefore the square root of the right-hand side:

$$\text{minimum error} = \sqrt{\left(\frac{\text{burette error}}{\text{burette volume}}\right)^2 + \left(\frac{\text{DVM error}}{\text{DVM reading}}\right)^2 + \left(\frac{\text{mass error}}{\text{overall mass}}\right)^2}$$

For this reason, we sometimes say the minimum error is the **root mean square (rms)** of the individual errors involved.

The burette: The error represents half the scale on the burette, 0.05 cm^3.
The DVM: The error represents half the scale on the DVM, 0.5 mV.
The top-pan balance: The error represents half the scale on the balance, 0.05 g.
Inserting values:

$$(\text{minimum error})^2 = \left(\frac{0.05 \text{ cm}^3}{1.0 \text{ cm}^3}\right)^2 + \left(\frac{0.5 \text{ mV}}{255 \text{ mV}}\right)^2 + \left(\frac{0.05 \text{ g}}{0.745 \text{ g}}\right)^2$$

....
Notice how the each of
the units cancel.
....

$$(\text{minimum error})^2 = (0.05)^2 + (1.96 \times 10^{-4})^2 + (0.0571)^2$$

$$(\text{minimum error})^2 = 2.5 \times 10^{-3} + 3.84 \times 10^{-8} + 4.5 \times 10^{-3}$$

that is:

$$(\text{minimum error})^2 = 7.0 \times 10^{-3}$$

so:

$$\text{minimum error} = \sqrt{7.0 \times 10^{-3}} = 0.084$$

This value means the final value of solubility product will have an error of 8.4%, assuming each stage of the measurement is performed with care. The error will of course be much larger if the chemist fails to take proper precautions, and is careless.

The calculation shows that most of the error (about 55%) is associated with the initial measurement of weighing out the KCl. For this reason, the weighing stage ought to be performed with special care. The error associated with the *emf* is negligible compared with the other errors. ■

> **Self-test 2**
>
> Calculate the errors associated with the following measurements:
>
> **2.1** Measuring a length ℓ of 11.2 cm with a ruler whose divisions are ℓ mm apart.
>
> **2.2** Measuring a temperature T of a water bath with a mercury-in-glass thermometer: $T = 24.9°C$ and the innate error is 0.5°C.
>
> **2.3** Measuring an enthalpy of reaction ΔH: the temperature change was 0.7°C and measured with a thermometer with divisions of 0.1°C; and weighing a mass of chemical of 3.503 g on a balance with divisions of 0.001 g.
>
> **2.4** Measuring the extinction coefficient ε via the Beer–Lambert law:
>
> $$Abs = \varepsilon[C]l$$
>
> where ℓ is the path length and $[C]$ is the concentration. The mass of analyte was 0.504 g measured on a balance with divisions of 0.001 g; the volume of solvent was 100 cm^3 in a volumetric flask with an innate error of 0.5 cm^3; the path length l was 1.0 cm as determined with a ruler with 0.5 mm divisions; and the absorbance Abs was 0.74 as determined with an instrument with an innate error of 0.05.

Graphical treatment of errors: error bars

By the end of this section, you will learn:

- When plotting a graph, the reading is plotted as a point, and the errors are indicated by **error bars** either side of this point, indicating the *range* of each reading.
- Such error bars may be indicated on both x and y axes.
- The best straight line through the data must accommodate all the error bars.

All readings have an associated error, so each variable determined experimentally represents a *range* of values of the form 'reading $\pm$ error'. We must accommodate these errors each time we draw a graph. In practice, we do not draw merely a single point per reading, but indicate this range using **error bars**.

> ■ **Worked Example 22.7**
>
> A chemical kineticist determines a pseudo first-order rate constant k' as a function of concentration $[C]$. Each value of k' has an error associated with it, as follows:
>
> | $[C]$/mol dm^{-3} | 0.1 | 0.2 | 0.3 | 0.4 | 0.5 |
> |---|---|---|---|---|---|
> | k'/s^{-1} | 9.0 ± 4.0 | 13.0 ± 2.0 | 22.2 ± 4.5 | 27.2 ± 2.0 | 33.3 ± 4.2 |
>
> Plot these data with k' (as y) against $[C]$ (as x), and determine a value for k_2.

Strategy:

(i) Each value of k' is plotted in the usual way, and is indicated with a bold point, as ●.

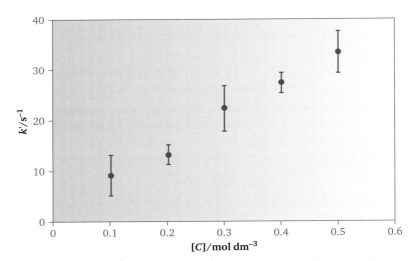

Figure 22.3 Plot of k' (as y) against $[C]$ (as x): the bold points • are the **readings** and the vertical lines joining the short horizontal lines—are the **error bars**.

(ii) The maximum and minimum values of k' are determined for each value of $[C]$, using the procedure demonstrated previously in Worked example 22.1. Such values are plotted on the same axes, with two short horizontal lines—positioned at equal distances above and below the respective values of $[C]$.

(iii) The **error bars** are formed by drawing a vertical line between the two horizontal lines—and passing through the points, •.

(iv) The second-order rate constant k_2 is obtained as the gradient of the graph.

Figure 22.3 shows a plot of k' (as y) against $[C]$ (as x). The graph clearly shows the relative magnitudes of the errors per point. The gradient of this graph is important, and yields the second-order rate constant k_2. Unfortunately, the data • clearly do not show a linear relationship between k' and $[C]$, making any determination of k_2 more difficult.

The best straight line through this data set must accommodate the errors (as indicated by the error bars ⌐) as well as the readings (the bold points, •). In practice, we draw two gradients on the graph, representing the steepest and the shallowest gradients possible. Each gradient must cut through all five of the error bars.

Figure 22.4 shows Fig. 22.3 redrawn with the lines of steepest and shallowest gradients superimposed. When we look at the steepest gradient (the dotted line ·····), we can see how the line only just touches the error bars for the first, fourth, and fifth points. If the dotted line were any steeper, it would fail to cut across one of more of these error bars; and if the line were less steep, its gradient would not be the maximum gradient.

In the same way, the solid line (——) on Fig. 22.4 only just cuts the error bars of the second and fourth points. If the line was steeper it would not have the minimum gradient; and if it were shallower, it would not cut across all five of the error bars.

We obtain a value for the second-order rate constant k_2 as the gradient of this figure; but the gradient has a range of values, as represented by the dotted and solid lines on Fig. 22.4. Accordingly, the value of k_2 must also have a range of values:

The **minimum** gradient is 51.0 $(\text{mol dm}^{-3})^{-1}\,\text{s}^{-1}$.

The **maximum** gradient is 81.3 $(\text{mol dm}^{-3})^{-1}\,\text{s}^{-1}$.

The maximum gradient is therefore nearly 60% larger than the minimum gradient.

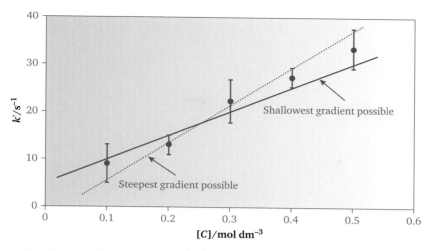

Figure 22.4 Plot of k' (as y) against $[C]$ (as x) showing error bars on the y axis: the dotted line has the maximum gradient possible with a line cutting all five error bars; the solid line has the minimum gradient possible.

We see how the value of k_2 lies in the range 51–81 $(mol \ dm^{-3})^{-1} \ s^{-1}$. ∎

Sometimes there is an error associated with the data in both the x and y axes, in which case we need two types of error bar. The error bars for the x axis are constructed in exactly the same way we constructed them for the y axis in Worked example 22.7, above.

■ **Worked Example 22.8**

An electrochemist wants to determine the change in entropy $\Delta S_{(cell)}$ associated with a cell discharging. To this end, we plot a graph of cell *emf* (as y) against temperature (as x). The gradient is termed the **voltage coefficient**. We calculate a value of $\Delta S_{(cell)}$ as 'voltage coefficient $\times n \ F$', where n is the number of electrons in the balanced cell reaction, and F is the Faraday constant.

   Using the data below, determine a range of values for $\Delta S_{(cell)}$ for this cell.

$\ominus Zn|ZnSO_{4(sat'd. \ soln.)}Hg_2SO_{4(s)}|Hg \oplus$

$T/°C$	20 ± 0.5	30 ± 0.5	40 ± 0.5	50 ± 0.5
emf/V	1.4430 ± 0.0004	1.4426 ± 0.0003	1.4421 ± 0.0003	1.4417 ± 0.0003

Figure 22.5 shows a graph of *emf* (as y) against temperature (as x), and containing error bars on both the x and y axes. The associated error of a point is contained within the bars. (For this reason, some people prefer mentally to superimpose an oval, centred on the data point.)

   The line of maximum gradient is shown as a dashed line, and the minimum gradient is drawn as a solid line:

   The **maximum** gradient is $4.67 \times 10^{-4} \ V \ K^{-1}$.

   The **minimum** gradient is $2.67 \times 10^{-5} \ V \ K^{-1}$.

In this example, the maximum gradient is nearly 18 times larger than the minimum gradient. ∎

The error in temperature is an innate error, i.e. using a relatively insensitive thermometer. The error in emf is probably due to fluctuations in the voltmeter reading.

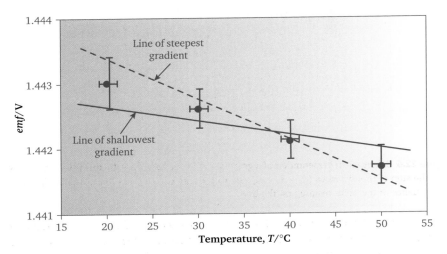

Figure 22.5 Plot of *emf* (as y) against T (as x), showing error bars on both x and y axes: the dotted line (– – – – – –) has the *maximum* gradient possible with a line cutting all five error bars; the solid line (————) has the *minimum* gradient possible.

In this last example, the spread of values is so large that we ought to stop and wonder: what then do we cite as 'the' gradient? We address this question in the next section.

Self test 3

The heat capacity C_p of chloroform (**I**) varies with temperature, according to the data below:

T/K	20 ± 0.5	30 ± 0.5	40 ± 0.5	50 ± 0.5
$C_p/J\ K^{-1}\ mol^{-1}$	91.47 ± 0.08	92.25 ± 0.07	93.02 ± 0.08	93.86 ± 0.06

Plot a graph of C_p (as y) against temperature T (as x), indicating all error bars.

I

The correlation coefficient *r*

By the end of this section, you will learn:

- The correlation coefficient *r* relates to the scatter of data.
- A correlation coefficient of 1 relates to a line with symmetrical scatter, with as many points on one side of the line as on the other.
- The value of *r* decreases as the scatter gets more random.

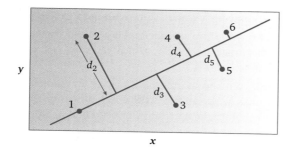

Figure 22.6 Schematic representation of a graph showing scatter. The solid line is the line of best fit if the sum of the distances $(d_2 + d_4 + d_6)$ equals the sum $(d_3 + d_5)$. In this example, $d_1 = 0$ in consequence of the point residing on the line.

There is no wholly safe way of deciding which is the best straight through a data set. Probably the simplest method is to consider the distances between the points on either side of the line, as follows.

Consider the schematic graph in Fig. 22.6, which clearly shows some scatter. The 'line of best fit' is drawn through the data, with points both above and below the line. We first describe in words why the line is 'the best'. The perpendicular distance between point 1 and the line is d_1, the perpendicular distance between point 2 and the line is d_2, etc. We would like the distances d_1 etc. to be as short as possible, but their sum Σd_n will depend on the extent of the scatter.

We now consider two sets of points: those above the line and those below. For each set, we can consider a sum of their distances from the line. We can call these two sums $\Sigma d_{n\,(above)}$ and $\Sigma d_{n(below)}$. If the line we drew through the data set is the *best* line, these two sums will be the same. The calculation of these sums is a straightforward process, although somewhat laborious.

> We define the **best** straight line through a data set as that for which the sum of the distances of the points *above* the line equals the corresponding sum for the points *below* it.

The **correlation coefficient** *r* is the most commonly used method of evaluating the scatter in a data set. In practice, it assesses whether a correlation exists between respective values of *x* and *y* within a data set. When the variables *x* and *y* are correlated, rather than following an explicit relationship, it is best to think in terms of a line of 'probable' fit rather than a line of 'best' fit. The closer the value of the observed variable to the probable value, the more likely is the relationship between *x* and *y*.

$$r = \frac{\sum_{i=1}^{i}[(x_i - \bar{x})(y_i - \bar{y})]}{\left\{\left[\sum_{i=1}^{i}(x_i - \bar{x})^2\right]\left[\sum_{i=1}^{i}(y_i - \bar{y})^2\right]\right\}^{1/2}} \qquad (22.5)$$

While this expression looks unpleasant, it is readily broken down into manageable chunks. All the terms have been defined previously.

The value of *r* can have any value between -1 and $+1$. A value of $r = +1$ indicates the data set does show a correlation, with all points lying on the straight line of positive gradient. A value of $r = -1$ defines a perfect negative correlation, i.e. a straight line with a negative slope. A value of *r* close to 0 indicates no correlation exists between respective values of *x* and *y* in the data set.

> The value of *r* obtained using eqn (22.5) is also called the **Pearson** correlation coefficient, and **the product moment** correlation coefficient.

We commonly cite a value of r^2 rather than *r* alone. This practice has two advantages. Firstly, it ensures values are always positive. Secondly, by squaring the value, we enhance the sensitivity of assessments based on the correlation coefficient; squaring accentuates the difference between *r* and the maximum value of 1.

We usually start by drawing a graph. Three situations can be anticipated:

1. **The graph is linear, with little scatter.** In this case, the correlation coefficient will probably lie in the range 0.90–0.95, which indicates we can safely assume a correlation does exist.

2. **The graph is linear, but shows extensive scatter.** In this case, we assume a poor correlation. This time, we might not wish to continue, obtain a duplicate set of data, choose to recalibrate the apparatus, or even find an alternative technique.

3. **The graph shows a smooth curve.** In this case, we assume a correlation exists, but not a *linear* correlation. We must first linearize the data (with the techniques in Chapter 12), and only then compute a correlation coefficient for the resulting straight line.

■ **Worked Example 22.9**

A graph is drawn of equilibrium constant K (as y) against temperature T (as x). The graph clearly shows a smooth curve. We suspect the data will follow the linear form of the van't Hoff isochore:

$$\ln K = -\frac{\Delta H^{\ominus}}{R}\frac{1}{T} + c$$

When calculating the correlation coefficient r, what parameter(s) should we include within eqn (22.5)?

The linear form of the van't Hoff isochore follows a line of the form $y = mx + c$, where $y = \ln K$, $m = -\Delta H \div R$, and $x = 1/T$. Accordingly:

Controlled variable	Observed variable
Symbol: x	Symbol: y
Choice of x: $1/T$	Choice of y: $\ln K$
Symbol for the mean of x: $\bar{x}$	Symbol for the mean of y: $\bar{y}$
Choice of $\bar{x}$: $\overline{T^{-1}}$	Choice of $\bar{y}$: $\overline{\ln K}$

A value of r^2 obtained before linearizing will be significantly less than 1; but, provided the data are good, correlation coefficient r^2 calculated with this choice of variables will have a value approaching 1. ■

Aside

These days, chemists generally obtain values of r using computer software. For example, using Excel ^{TM}. The procedure to obtain r with a graph is:

1. Draw a graph using Excel^{TM} in the usual way.

2. Place the mouse on one of the data points, and click once. This action will highlight the data.

3. With the mouse, click on the CHART menu, and choose ADD TRENDLINE.

4. Click on the top-left icon, which is labelled 'LINEAR'.

5. At the top of the box, now click on the second page, labelled 'OPTIONS'.

6. Click the box labelled 'DISPLAY R-SQUARED VALUE ON GRAPH'.

7. Click on OK.

A small text box will appear on the graph, citing the value of r^2.

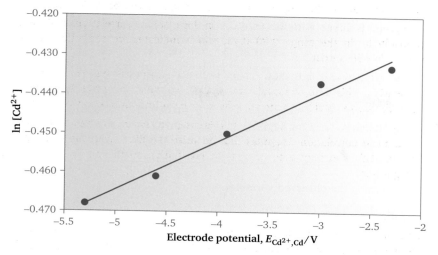

Figure 22.7 Nernst plot of $E_{Cd^{2+},Cd}$ (as y) against ln $[Cd^{2+}]$ (as x). We obtain $E^{\ominus}_{Cd^{2+},Cd}$ as the intercept on the y axis.

■ **Worked Example 22.10**

An electrochemist determines values of the electrode potential $E_{Cd^{2+},Cd}$ as a function of cadmium-ion concentration, $[Cd^{2+}]$, as follows:

$[Cd^{2+}]$/mol dm^{-3}	0.1	0.05	0.02	0.01	0.005
$E_{Cd^{2+},Cd}$/V	-0.433	-0.437	-0.450	-0.461	-0.468

The electrochemist seeks to determine a value of the standard electrode potential $E^{\ominus}_{Cd^{2+},Cd}$, so plots a graph of ln $[Cd^{2+}]$ (as y) against $E_{Cd^{2+},Cd}$ (as x). Calculate the correlation coefficient, *r*.

Figure 22.7 shows a graph of ln $[Cd^{2+}]$ (as y) against $E_{Cd^{2+},Cd}$ (as x). The graph is clearly linear, but has some slight scatter. The line of best fit has been drawn already.

Strategy:

(i) Calculate the mean of both x and y (in this case, the means of $E_{Cd^{2+},Cd}$ and ln $[Cd^{2+}]$, respectively).

(ii) Calculate the terms $(x_i - \bar{x})$ and $y_i - \bar{y}$, and hence determine $\sum_{i=1}^{i}[(x_i - \bar{x}) \times (y_i - \bar{y})]$.

(iii) Calculate the terms $(x_i - \bar{x})^2$ and $(y_i - \bar{y})^2$, and hence determine values of
$$\sum_{i=1}^{i}(x_i - \bar{x})^2, \sum_{i=1}^{i}(y_i - \bar{y})^2, \text{ and then } \left\{ \left[\sum_{i=1}^{i}(x_i - \bar{x})^2\right]\left[\sum_{i=1}^{i}(y_i - \bar{y})^2\right] \right\}^{1/2}.$$

(iv) Insert the terms from parts (i) and (ii) into eqn (22.5).

Solution

(i) The mean of ln $[Cd^{2+}] = \bar{x} = 3.823$, and the mean of $E_{Cd^{2+},Cd} = \bar{y} = 0.4498$ V.

> We sometimes refer to this set of axes as a **Nernst plot**. We obtain $E^{\ominus}_{Cd^{2+},Cd}$ as the intercept on the y axis.

(ii) Concerning the **controlled variable**:

[Cd^{2+}]	$x = \ln$ [Cd^{2+}]	$(x_i - \bar{x})$	$(x_i - \bar{x})^2$
0.1	−2.303	1.520	2.311
0.05	−2.996	0.827	0.684
0.02	−3.912	−0.089	0.008
0.01	−4.605	−0.782	0.612
0.005	−5.298	−1.476	2.177

Concerning the **observed variable**:

$y = E_{Cd^{2+},Cd}$	$(y_i - \bar{y})$	$(y_i - \bar{y})^2$
−0.433	0.017	2.822×10^{-4}
−0.437	0.013	1.638×10^{-4}
−0.450	0.000	9.000×10^{-4}
−0.461	−0.011	1.254×10^{-4}
−0.468	−0.018	3.312×10^{-4}

Working with data from the penultimate columns in the two tables, we calculate:

$$\sum_{i=1}^{i} [(x_i - \bar{x})(y_i - \bar{y})]$$

So the numerator of eqn (22.5) = 0.0712.

(iii) Working with data in the far right-hand columns in the two tables, we calculate:

$$\sum_{i=1}^{i} (x_i - \bar{x})^2 = 5.792$$

$$\sum_{i=1}^{i} (y_i - \bar{y})^2 = 9.029 \times 10^{-4}$$

Therefore:

$$\left\{ \left[\sum_{i=1}^{i} (x_i - \bar{x})^2 \right] \left[\sum_{i=1}^{i} (y_i - \bar{y})^2 \right] \right\}^{1/2} = \sqrt{(5.792) \times (9.029 \times 10^{-4})}$$

So the denominator of eqn (22.5) = $\sqrt{5.2296 \times 10^{-3}}$ = 0.0723.

(iv) Finally, we insert values into eqn (22.5):

$$r = \frac{0.0712}{0.0723} = 0.9848$$

In summary, $r = 0.9848$, so $r^2 = 0.9698$.

This value looks reasonable because it is close to unity; and the data show a clear correlation, though with some slight scatter. ∎

Remember: The **numerator** is the top line of a quotient or fraction The **denominator** is the bottom line of a quotient or fraction

Self-test 4

The freezing temperature of water depends on its purity, as expressed by the concentration of sodium chloride it contains.

The following melting temperatures were measured:

[NaCl]/mol dm^{-3}	0	0.1	0.2	0.3	0.4	0.5
$T_{(freeze)}/K$	273.15	273.01	272.94	272.88	272.80	272.72

Determine a value for the correlation coefficient r for this data set.

Additional problems

22.1 A voltmeter fluctuates when the analyte solution is stirred. When the *emf* is 0.340 V, the maximum fluctuation is 10 mV. What is the signal-to-noise ratio?

22.2 The progress of a simple kinetics experiment is followed with a normal wristwatch. The watch has the three usual hands of hours, minutes, and seconds, and the watch face has a printed mark for every second. Determine the innate error in the time measurement.

22.3 Consider the data below, which relate to the second-order racemization of glucose (**II**) in aqueous hydrochloric acid at 17 °C. The concentrations of glucose and hydrochloric acid are the same, [A]. The concentration $[A]_t$ decreases with time according to the linear form of the integrated second-order rate equation:

$$\frac{1}{[A]_t} = kt + c$$

Time, t/s	0	600	1200	1800	2400
$[A]_t$/mol dm^{-3}	0.400	0.360	0.301	0.289	0.244

Linearize the data and plot a graph. What is the correlation coefficient, r?

II

22.4 The optical absorbance of a dye in solution is 3.2, but the amplitude of the interference is 0.7. Calculate the signal-to-noise ratio.

22.5 The innate error in the absorbance reading with a modern spectrophotometer can be as low as 0.001. If the absorbance is 0.670, calculate the minimum error.

22.6 A chemist determines a rate constant by removing an aliquot of solution every 10 minutes, and titrating it with base. Calculate the minimum error if:

- The aliquot volume is $10.0\,cm^3$, removed with a pipette with an innate error of $0.005\,cm^3$.
- The concentration of base is $0.104 \pm 0.001\,moldm^{-3}$.
- Because the solution is not removed instantaneously, but has to fill the pipette, the innate error in the time reading is 15 s.

22.7 Consider the isomerization of 1-butene (**III**) to form *trans* 2-butene (**IV**). The equilibrium constants of reaction are given in the table below:

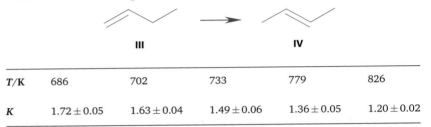

T/K	686	702	733	779	826
K	1.72 ± 0.05	1.63 ± 0.04	1.49 ± 0.06	1.36 ± 0.05	1.20 ± 0.02

Plot a graph of ln K (as y) against $1/T$ (as x). Determine the maximum and minimum gradients possible.

22.8 A cell is constructed to help determine the standard electrode potential of the copper(II)–copper couple, and using a variant of the Nernst equation:

$$emf = \left(E^{\ominus}_{Cu^{2+},Cu} + \frac{RT}{2F}\ln[Cu^{2+}] \right) - E_{SCE}$$

The *emf* is 28.3 mV, with an innate error of 1 mV. The concentration $[Cu^{2+}]$ is 4.4×10^{-3} mol dm^{-3}, with an innate error of 1×10^{-5} mol dm^{-3}. What is the minimum error associated with this measurement?

22.9 Return to the graph in Self-test 3, and determine the maximum and minimum gradients.

22.10 Return to question 22.7 above. What is the correlation coefficient, *r*? (Ignore the error bars.)

Answers to self-test questions

Chapter 1:
The display of numbers: standard factors, scientific notation, and significant figures

Self-test 1

	variable	number	factor	unit(s)
1.1	mass	2.65	k (10^3)	g
1.2	frequency	94.5	M (10^6)	Hz
1.3	wavelength	500	n (10^{-9})	m
1.4	current	0.3	m (10^{-3})	A
1.5	mass of beaker	340	(none)	g
1.6	mass of large car	1.4	M (10^6)	g
1.7	energy liberated	34	M (10^6)	J mol^{-1}
1.8	length of bond	130	p (10^{-12})	m
1.9	potential difference	550	m (10^{-3})	V
1.10	speed	34	(none)	m s^{-1}

Self-test 2

2.1   energy $= 12.3\,\text{kJ mol}^{-1}$

2.2   frequency $= 500\,\text{MHz}$ or $0.5\,\text{GHz}$

2.3   length of road $= 3.4\,\text{km}$

2.4   voltage $= 30\,\text{kV}$

2.5   mass of truck $= 36\,\text{Mg}$ (which, incidentally, is the same as 3.6 tonnes)

2.6   cost $= £1.2\,\text{M}$

2.7   energetic content $= 2.034\,\text{MJ}$

2.8   wavelength $\lambda = 0.98\,\mu\text{m}$ or $980\,\text{nm}$

2.9   bond length $= 156\,\text{pm}$

2.10   diameter of a fly's eye $= 10\,\mu\text{m}$

Self-test 3

3.1   energy $= 12.3 \times 10^3\,\text{J mol}^{-1}$

3.2   frequency $= 500 \times 10^6\,\text{Hz}$ or $0.5 \times 10^9\,\text{Hz}$

3.3   length of road $= 3.4 \times 10^3\,\text{m}$

3.4   voltage $= 30 \times 10^3\,\text{V}$

3.5   mass of truck $= 36 \times 10^6\,\text{g}$

3.6   cost $= £1.2 \times 10^6$

3.7   energetic content $= 2.\,034 \times 10^6\,\text{J}$

3.8   wavelength $\lambda = 0.98 \times 10^{-6}\,\text{m}$ or $980 \times 10^{-9}\,\text{m}$

3.9   bond length $= 156 \times 10^{-12}\,\text{m}$

3.10   diameter of a fly's eye $= 10 \times 10^{-6}\,\text{m}$

3.11 $F = 9.65 \times 10^5 \, C \, mol^{-1}$

3.12 $q = 1.6 \times 10^{-19} \, C$

Self-test 4

4.1 energy change $= -135 \, kJ$

4.2 $emf = 1.432 \, V$

4.3 volume $= 2.0 \, m^3$

4.4 amount of material $= 3.22 \, mol$

4.5 mass $= 1 \, Mg$

4.6 1 meter $= 1 \, 650 \, 764$ wavelengths

Self-test 5

5.1 $V = 120 \times 151 \times 146.5 = 2 \, 654 \, 580 \, pm^3$. Because a is expressed to two s.f.s, $V = 2.7 \times 10^6 \, pm^3$.

5.2 $n = 0.250 \times 0.05 \, mol = 0.0125 \, mol$. Because the concentration was expressed to one s.f., $n = 0.01 \, mol$.

5.3 rate $= 9.3 \times 10^{-2} \times 0.3 \, mol \, dm^{-3} \, s^{-1} = 0.0279 \, mol \, dm^{-3} \, s^{-1}$. Because the concentration was expressed to one s.f., rate $= 3 \times 10^{-2} \, mol \, dm^{-3} \, s^{-1}$.

5.4 $c = 3.2 \times 10^{-4} \div 0.5 \, mol \, dm^{-3} = 6.4 \times 10^{-4} \, mol \, dm^{-3}$. Because the volume was expressed to one s.f., $c = 6 \times 10^{-4} \, mol \, dm^{-3}$.

5.5 $n = 4 \, g \div 422 \, g \, mol^{-1} = 9.4797 \times 10^{-4} \, mol$. Because the mass was expressed to one s.f., $n = 1 \times 10^{-3} \, mol = 1 \, mmol$.

5.6 $Q = 87.3 \, mC \div 0.32 \, cm^2 = 0.272 \, 812 \, 5 \, C \, cm^2$. Because the area was expressed to two s.f.s, $Q = 0.27 \, C \, cm^2$.

5.7 $p = \dfrac{0.13 \times 8.314 \times 298}{14.2} \, Pa = 22.681 \, 997 \, Pa$. Because the amount of gas n was expressed to expressed to two s.f.s, $p = 23 \, Pa$.

5.8 $\Delta G^{\ominus} = -8.314 \, J \, K^{-1} \, mol^{-1} \times 298 \, K \times 4.0 = -2477.572 \, J \, mol^{-1} \times 4.0 = -9972.9706 \, J \, mol^{-1}$. Because K was expressed to two s.f.s, $\Delta G^{\ominus} = -2.0 \, kJ \, mol^{-1}$.

Self-test 6

6.1 Rewriting: mass $= (12.0 + 1001 - 130.62) \, g = 1042.62 \, g$. This is 1043 g to zero d.p.s.

6.2 Charge $= 94 \, 836.43 \, C$. This is 94 836 C to zero d.p.s.

6.3 Amount of material $= 9.8879 \, mol$. This is 9.9 mol to one d.p.

6.4 Rewriting: time $= 60.4 \, s + 0.000 \, 001 \, 2 \, s + 0.033 \, 96 \, s + 4.0 \, s = 64.433 \, 961 \, 2 \, s$. This is 64.4 s to one d.p.

Chapter 2:
Algebra I: introducing notation, nomenclature, symbols, and operators

Self-test 1

1.1 $\Delta X = 10 - 22 = -12$

1.2 $\Delta X = 33 - 5.2 = 27.8$

1.3 $\Delta X = 9.37 - 9.34 = 0.03$

1.4 $\Delta G = 8 - 3.6 \, kJ \, mol^{-1} = 4.4 \, kJ \, mol^{-1}$

1.5 $\Delta(Abs) = 0.6 - 1.1 = -0.5$

1.6 $\Delta(emf) = 1.50 - 1.45 \, V = 0.05 \, V$

Self-test 2

2.1 $422 \, g \, mol^{-1}$

2.2 $160 \, g \, mol^{-1}$

2.3 $34 \, g \, mol^{-1}$

2.4 $122\,\text{g}\,\text{mol}^{-1}$
2.5 $79\,\text{g}\,\text{mol}^{-1}$
2.6 $180\,\text{g}\,\text{mol}^{-1}$

Self-test 3

3.1   Scheme 1.1: Compound (**IX**) has a molar mass of $58\,\text{g}\,\text{mol}^{-1}$ and compound (**X**) has a molar mass of $62\,\text{g}\,\text{mol}^{-1}$. Water has a molar mass of $18\,\text{g}\,\text{mol}^{-1}$. Therefore:

$$\text{Mass of product} = \sum(0.35 \times 58) + (0.35 \times 62) - (0.35 \times 18)\,\text{g}$$
$$\text{Mass of product} = \sum(20.3 + 21.7 - 6.3) = 35.7\,\text{g}$$

3.2   Each component within the sum is (credits × percentage). We write:

$$\sum[(10 \times 45) + (20 \times 56) + (20 \times 70) + (20 \times 62) + (30 \times 40)]$$

$$\begin{array}{ccccc}\textbf{maths} & \textbf{inorganic} & \textbf{organic} & \textbf{physical} & \textbf{laboratory}\end{array}$$
$$\sum 450 + 1120 + 1400 + 1240 + 1200 = 5410$$

The total number of marks the student could have obtained in 100 credits, each of 100%, is 10 000.
Accordingly, the student's final percentage mark is: $(5410/10\,000) \times 100 = 54.1.\%$

Self-test 4

4.1 $1 \times 2 \times 3 \times 4 \times 5 = 120$
4.2 $1/2^2 \times 1/3^2 \times 1/4^2 = 1/4 \times 1/9 \times 1/16 = 1/(4 \times 9 \times 16) = 1/576 = 0.001\,736$
4.3 $5^3 \times 6^3 \times 7^3 = 125 \times 216 \times 334 = 9\,261\,000$
4.4 $\sqrt{1} \times \sqrt{2} \times \sqrt{3} = (1) \times (1.414) \times (1.732) = 2.45$
4.5 $\displaystyle\prod_{i=5}^{10} i$ 4.6 $\displaystyle\prod_{i=9}^{12} 1/i$

4.7 $\displaystyle\prod_{i=3}^{6} i^2$ 4.8 $\displaystyle\prod_{i=1}^{5} i^2$

Chapter 3:
Algebra II: the correct order to perform a series of operations: BODMAS

Self-test 1

1.1 $x = 36$ (the calculation is *associative*, so the order does not matter)
1.2 $x = (6 \times 7) - (8 \times 2) = 42 - 16 = 26$ 1.3 $x = (2+3) \times 5 = (5) \times 5 = 25$

1.4 $x = 4 + 25 - 81 = -52$ 1.5 $x = \dfrac{24}{14} = 1.714$

1.6 $x = \dfrac{(54) - 2}{3 + (10)} = \dfrac{52}{13} = 4$

1.7 $x = 5 \times \sqrt{4 - (12) + 44} = 5 \times \sqrt{36} = 5 \times 6 = 30$

1.8 $x = \left(\dfrac{12}{6}\right)^2 - 56 = (2)^2 - 56 = 4 - 56 = -52$

Chapter 4:
Algebra III: simplification and elementary rearrangements

Self-test 1

1.1 $7a$
1.2 $33b$
1.3 $23c$

1.4 $12g + 3h + 7i$

1.5 $16g$

1.6 $14f + 8e - 7d$

1.7 $0f - h$ (which is the same as $-1 \times h + 10f$)

1.8 $4C + 10H + S + O$ (which is the same as the empirical formula, $C_4H_{10}SO$)

Self-test 2

2.1   9 has been ADDED to the left-hand side, so we so we perform the inverse operation and SUBTRACT 9 from both sides: $x = 12 - 9 = 3$.

2.2   We first collect terms: $x - 4v = 4v$. Next, since $4v$ has been SUBTRACTED from the left-hand side, we perform the inverse operation and ADD $4v$ to both sides: $x = 4v + 4v$. We then tidy up by collecting terms: $x = 8v$.

2.3   Because 12 has been SUBTRACTED from x, we perform the inverse operation, and ADD 12 to both sides: so $a + 12 = x$.

2.4 $4c$ has been ADDED to x, so we perform the inverse operation, and SUBTRACT $4c$ from both sides: so $a - 4c = x$.

2.5 $m_{(two)}$ has been SUBTRACTED from $m_{(one)}$, so we perform the inverse operation and ADD $m_{(two)}$ to both sides: so, $4p + m_{(two)} = m_{(one)}$.

2.6   We first group terms: $p = x + 5b$. $5b$ has been ADDED to x, so we SUBTRACT $5b$ from both sides: so, $p - 5b = x$.

Self-test 3

3.1 $18 \div 6 = 3$, so $y = 3c$

3.2 $2 \div 4 = \dfrac{1}{2}$, so $y = \dfrac{d}{2}$ $\left(\text{or } y = \dfrac{1}{2}d \right)$

3.3 $d \div d = 1$, so $y = \dfrac{1}{4}$

3.4 c and d both appear on top and bottom, so $y = b/e \times c/c \times d/d$, which yields, $y = b/e$.

3.5   Rewriting slightly: $y = \dfrac{6}{3d} \times \dfrac{4}{20} = \dfrac{2}{d} \times \dfrac{1}{5}$, so $= \dfrac{2}{5d}$.

3.6 $y = (6a \times a \times b)/6a$. The $6a$ terms on top and bottom cancel, leaving $y = ab$.

Self-test 4

4.1 $x = y/m$

4.2 $x = 4/12 = 1/3$

4.3 $240x/240 = 50y/240$ so $x = 5y/24$

4.4 $x = 12/ab$

4.5 $x = mgh/55$

4.6 $x = 55h/mg$

4.7 $11x = g$, so $x = g/11$

4.8 $8x = 34$, so $x = 34/8$, and $x = 17/4$

4.9 $6f = 13x$, so $x = 6f/13$

4.10 $p - 2q = 6x$, so $x = (p - 2q)/6$

Self-test 5

5.1 $y = 3x^2 \quad \rightarrow \quad y/3 = x^2 \quad \rightarrow \quad x = \sqrt{y/3}$

5.2 $4y = ax^3 \quad \rightarrow \quad 4y/a = x^3 \quad \rightarrow \quad x = \sqrt[3]{4y/a}$

5.3 $by = ax^4 \quad \rightarrow \quad by/a = x^4 \quad \rightarrow \quad x = \sqrt[4]{by/4}$

5.4 $y = \sqrt{x} \quad \rightarrow \quad x = y^2$

5.5 $ay = 4\sqrt{x} \quad \rightarrow \quad ay/4 = \sqrt{x} \quad \rightarrow \quad x = (ay/4)^2$

5.6 $y = \sqrt[4]{x} \quad \rightarrow \quad x = y^4$

5.7 $\delta = \dfrac{1.61v^{1/6} D^{1/3}}{\sqrt{\omega}} \quad \rightarrow \quad \delta\sqrt{\omega} = 1.61v^{1/6}D^{1/3}$

        $\rightarrow \quad \sqrt{\omega} = \dfrac{1.61v^{1/6} D^{1/3}}{\delta} \quad \rightarrow \quad \omega = \left(\dfrac{1.61\, v^{1/6} D^{1/3}}{\delta} \right)^2$

Chapter 5:
Algebra IV: rearranging equations according to the rules of algebra

Self-test 1

1.1 Combining terms: $y = 8x + 1$. SUBTRACTING the 1: $y - 1 = 8x$. DIVIDING by 8: $x = \dfrac{y - 1}{8}$.

1.2 $y = 3x - 7 \quad \rightarrow \quad y + 7 = 3x \quad \rightarrow \quad x = \dfrac{y + 7}{3}$

1.3 $y = 5 - 4x \quad \rightarrow \quad y - 5 = -4x \quad \rightarrow \quad x = \dfrac{y - 5}{-4} = \dfrac{5 - y}{4}$

1.4 $y = \dfrac{x - 4}{7} \quad \rightarrow \quad 7y = x - 4 \quad \rightarrow \quad x = 7y + 4$

1.5 $y = 8(x - 1) \quad \rightarrow \quad \dfrac{y}{8} = x - 1 \quad \rightarrow \quad x = \dfrac{y}{8} + 1$

1.6 $y = 8x(b - 1) \quad \rightarrow \quad \dfrac{y}{(b - 1)} = 8x \quad \rightarrow \quad x = \dfrac{y}{8(b - 1)}$

In fact, we could have performed these two steps at the *same* time.

1.7 $y = \dfrac{d}{x} \quad \rightarrow \quad xy = d \quad \rightarrow \quad x = \dfrac{d}{y}$

1.8 $y = \dfrac{3}{x - 2} \quad \rightarrow \quad y(x - 2) = 3 \quad \rightarrow \quad \dfrac{3}{y} = x - 2 \quad \rightarrow \quad x = \dfrac{3}{y} + 2$

1.9 $y = \dfrac{a - x}{4d} \quad \rightarrow \quad 4dy = a - x \quad \rightarrow \quad 4dy - a = -x \quad \rightarrow \quad x = a - 4dy$

The last step represents MULTIPLYING both sides of the equation by -1.

1.10 $y = \dfrac{x - 9}{c + d} \quad \rightarrow \quad y(c + d) = x - 9 \quad \rightarrow \quad x = y(c + d) + 9$

1.11 $C = \dfrac{5}{9} \times F - 32 \quad \rightarrow \quad C + 32 = \dfrac{5}{9} \times F \quad \rightarrow \quad 9(C + 32) = 5F \quad \rightarrow$
$F = \dfrac{9(C + 32)}{5}$

1.12 $c^2 = a^2 + b^2 \quad \rightarrow \quad c^2 - a^2 = b^2 \quad \rightarrow \quad b = \sqrt{c^2 - a^2}$

Self-test 2

2.1 $y = x^2 \quad \rightarrow \quad x = \sqrt{y}$

2.2 $y = -4x^2 \quad \rightarrow \quad -\dfrac{y}{4} = x^2 \quad \rightarrow \quad x = \sqrt{-\dfrac{y}{4}}$

2.3 $y = x^3 + 7 \quad \rightarrow \quad y - 7 = x^2 \quad \rightarrow \quad x = \sqrt{y - 7}$

2.4 $y = 4(c - x^2) \quad \rightarrow \quad \dfrac{y}{4} = c - x^2 \quad \rightarrow \quad \dfrac{y}{4} + x^2 = c \quad \rightarrow \quad x^2 = c - \dfrac{y}{4} \quad \rightarrow$
$x = \sqrt{c - \dfrac{y}{4}}$

2.5 $y = c(x^2 + 1) \quad \rightarrow \quad \dfrac{y}{c} = x^2 + 1 \quad \rightarrow \quad \dfrac{y}{c} - 1 = x^2 \quad \rightarrow \quad x = \sqrt{\dfrac{y}{c} - 1}$

2.6 $y = (x - a)^2 \quad \rightarrow \quad \sqrt{y} = x - a \quad \rightarrow \quad x = \sqrt{y} + a$

2.7 $y = \sqrt{x - 9} \quad \rightarrow \quad y^2 = x - 9 \quad \rightarrow \quad x = y^2 + 9$

2.8 $y = 5 \times \sqrt{x - v} \quad \rightarrow \quad \dfrac{y}{5} = \sqrt{x - v} \quad \rightarrow \quad \left(\dfrac{y}{5}\right)^2 = x - v \quad \rightarrow \quad x = \left(\dfrac{y}{5}\right)^2 + v$

2.9 $y = \sqrt{a - x} \quad \rightarrow \quad y^2 = a - x \quad \rightarrow \quad x + y^2 = a \quad \rightarrow \quad x = a - y^2$

2.10 $y = \left(\dfrac{x + 1}{5}\right)^2 \quad \rightarrow \quad \sqrt{y} = \dfrac{x + 1}{5} \quad \rightarrow \quad 5\sqrt{y} = x + 1 \quad \rightarrow \quad x = 5\sqrt{y} - 1$

Chapter 6:
Algebra V: simplifying equations: brackets and factorizing

Self-test 1

1.1 $pa + pb$

1.2 $bpx - bpy$

1.3 $ac + 2a + c + 2$

1.4 $a^2 \times a = a^3$, so $a^3 - ab + a^2b^2 - b^3$

1.5 It is easier if the MULTIPLICATION process is performed in two separate stages, so:
$g \times [HJ + 3H + 2J + 6]$, then $gHJ + 3gH + 2gJ + 6g$

1.6 $J^2 - 4$

1.7 $3/2 - a/2c + 12v - 4va/c$

1.8 $e/c + 5/c - e/d - 5/d$

Self-test 2

Remember that $'-' \times '-' = '+'$.

2.1 $a^2 + 2ab + b^2$

2.2 $x^2 - 2xy + y^2$

2.3 $x^2 - 2xc + c^2$

2.4 $x^2 + 2xy + y^2$

2.5 $4x^2 - 4xy + y^2$

2.6 $a^2x^2 - 2abxy + b^2y^2$

2.7 $25 - 10y^2 + y^4$

2.8 $1 - 8y + 16y^2$

Self-test 3

3.1 $a(1 + b)$

3.2 $2(x - y)$

3.3 $a(b + 2c + d)$

3.4 $2a(2 + c + 3b)$

3.5 $a(a + b)$

3.6 $a(a + 4)$

3.7 $a(a + 3 + 6b)$

3.8 $x(y - x + 2)$

3.9 $x(x^2 + 6x + 4)$

3.10 (i) We multiply out the brackets: $x^2 + 8x + 16 + 4x^2 - 12x + 9 - 25 - x$.
(ii) We collect together the like terms: $5x^2 - 5x$.
(iii) We factorize: $5x(x - 1)$.

Self-test 4

4.1 $(x - b)(x + b)$

4.2 $(c - p)(c + p)$

4.3 $(\alpha - \beta)(\alpha + \beta)$

4.4 $(\omega - \pi)(\omega + \pi)$

4.5 $(x - 2)(x + 2)$

4.6 $(a - 9)(a + 9)$

4.7 $(d - 100)(d + 100)$

4.8 $\sqrt{5.32} = 2.31$, so $(a - 2.31)(a + 2.31)$

Self-test 5

5.1 $(x + 3)(x + 1)$

5.2 $(x + 2)(x + 3)$

5.3 $(x + 4)(x + 1)$

5.4 $(x + 6)(x + 1)$

5.5 $(x + 4)(x + 5)$

5.6 $(x + 20)(x + 5)$

5.7 $(x + 3)(x + 3)$

5.8 $(x+8)(x+7)$
5.9 $(x+2)(x+5)$
5.10 $(x+3)(x+5)$

Self-test 6

6.1 $(x+4)(2x-3)$
6.2 $(2x+2)(2x+3)$
6.3 $(x+1)(10x-1)$
6.4 $(x-2)(5x-1)$
6.5 $(x-0.5)\left(\dfrac{1}{2}x+2\right)$
6.6 $(x+1)(2x-0.4)$
6.7 $(x+0.123)(x+0.456)$
6.8 $(x+1.31)(x-3.81)$

Chapter 7:
Graphs I: introducing pictorial representations of functions

Self-test 1

1.1   Quantitative
1.2   Qualitative
1.3   Qualitative
1.4   Quantitative
1.5   Qualitative (i.e. later than we should be but by an unspecified amount of time)
1.6   Quantitative

Self-test 2

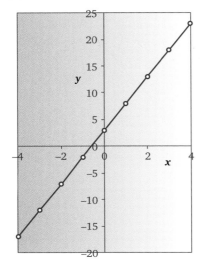

2.1

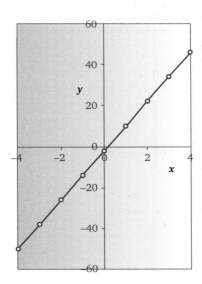

2.2

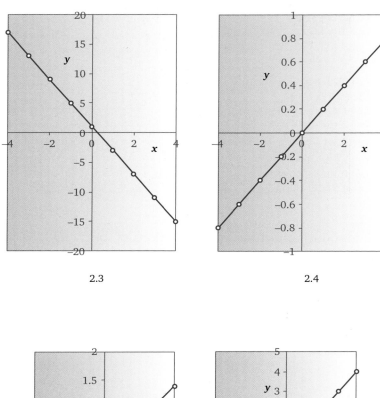

2.3

2.4

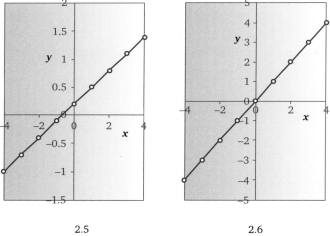

2.5

2.6

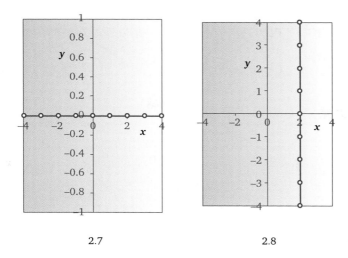

2.7

2.8

Self-test 3

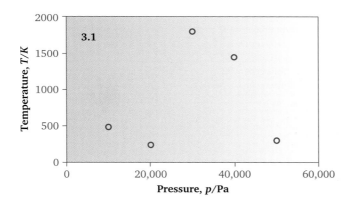

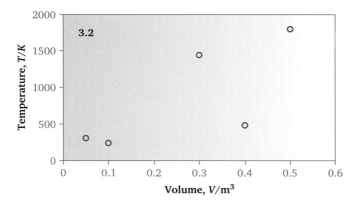

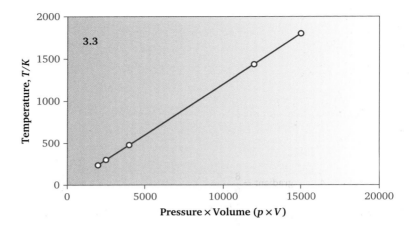

Chapter 8:
Graphs II: the equation of a straight-line graph

Self-test 1

1.1   gradient $= 2$ intercept $= 4$
1.2   gradient $= 3.5$ intercept $= -2$
1.3   gradient $= 4$ intercept $= 8.4$
1.4   gradient $= 4$ intercept $= 3$
1.5   gradient $= 1$ intercept $= 22$
1.6   gradient $= -4$ intercept $= -4.3$

Self-test 2

2.1 $y = 2x + 5$
2.2 $y = -(5/3)x - 10$
2.3 $y = -2x - 3$
2.4 $y = (4/7.1)x$ (which could be written as $y = (4/7.1)x + 0$)
2.5 $y = (19/30.2)x + 10.4/30.2$
2.6 $y = x - 3$
2.7 $y = -(2/3.17)x + 1.22/3.17$
2.8 $y = 1/10^2 x - 10^3$

Self-test 3

3.1   3
3.2   0
3.3 -0.5

Self-test 4

4.1 $\dfrac{4 - 2}{2 - 1} = \dfrac{2}{1} = 2$

4.2 $\dfrac{22 - 17}{5 - 10} = \dfrac{5}{-5} = -1$

4.3 $\dfrac{14 - 5}{6 - 3} = \dfrac{9}{3} = 3$

4.4 $\dfrac{7-3}{0-(-2)} = \dfrac{4}{2} = 2$

4.5 $\dfrac{-5-(-3)}{-5-(-3)} = \dfrac{-2}{-2} = 1$

4.6 $\dfrac{5-(-3)}{4-2} = \dfrac{8}{2} = 4$

4.7 $\dfrac{-4-(-8)}{-4-(-2)} = \dfrac{4}{-2} = -2$

4.8 $\dfrac{12-4}{7-(-5)} = \dfrac{8}{12} = \dfrac{3}{4}$

Self-test 5

5.1 intercept $= 4$ gradient $= \dfrac{8-4}{6-0} = \dfrac{2}{3}$ line $= y = \dfrac{2}{3}x + 4$

5.2 intercept $= 6$ gradient $= \dfrac{12-6}{30-0} = \dfrac{6}{30} = \dfrac{1}{5}$, or 0.2 line $= y = 0.2x + 6$

5.3 intercept $= 40$ gradient $= \dfrac{40-10}{0-15} = \dfrac{30}{-15} = -2$ line $= y = -2x + 40$

Self-test 6

6.1 $y = 3x - 1$
6.2 $y = 10x - 36$
6.3 $y = -4x + 12$
6.4 $y = 2.5x - 5$
6.5 $y = -3.5x - 14$

Self-test 7

7.1 gradient $= \dfrac{4-2}{2-1} = 2$ line $= y = 2x + 0$ i.e. $y = 2x$

7.2 gradient $= \dfrac{-11-(-2)}{3-0} = \dfrac{-9}{3} = -3$ line $= y = -3x - 2$

7.3 gradient $= \dfrac{50-12}{26-9} = \dfrac{38}{19} = 2$ line $= y = 2x - 6$

7.4 gradient $= \dfrac{10-9}{6-7} = \dfrac{1}{-1} = -1$ line $= y = -x + 16$

7.5 gradient $= \dfrac{-4-2}{-3-1} = \dfrac{-6}{-4} = \dfrac{3}{2}$ line $= y = \dfrac{3}{2}x + \dfrac{1}{2}$

7.6 gradient $= \dfrac{-6-(-3)}{-2-(-3)} = \dfrac{-3}{1} = -3$ line $= y = -3x - 12$

Chapter 9:
Graphs III: straight lines that intersect

Self-test 1

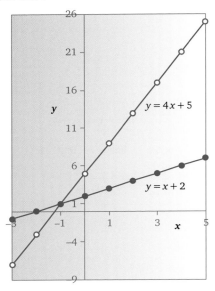

1.1   The lines intersect at $(-1,1)$.

1.2   The lines intersect at $(2,0)$.

Self-test 2

2.1 $x=1$ and $y=7$
2.2 $x=-4$ and $y=-2$
2.3 $x=-1$ and $y=3$
2.4 $x=0$ and $y=5$
2.5 $x=0$ and $y=3$
2.6 $x=-3$ and $y=-2$

Self-test 3

3.1 $x=8$ and $y=-21$
3.2 $x=-4$ and $y=-2$
3.3 $x=1$ and $y=2$
3.4 $x=3$ and $y=4$
3.5 $x=2$ and $y=3$
3.6 $x=7$ and $y=9$
3.7 $x=11$ and $y=5$
3.8 $x=2.5$ and $y=3$

Self-test 4

4.1 $x=3$ and $y=4$
4.2 $x=2$ and $y=5$
4.3 $x=3$ and $y=14$
4.4 $x=2$ and $y=-1$
4.5 $x=0.5$ and $y=1$
4.6 $x=-0.4$ and $y=12$

Chapter 10:
Powers I: introducing indices and powers

Self-test 1

1.1 T^7
1.2 c^5
1.3 a^3
1.4 q^9
1.5 f^1
1.6 g^5

Self-test 2

2.1 h^5g^2
2.2 $c^2d^2g^2h^6$
2.3 $c^1d^1e^1f^2$
2.4 $x^4h^2l^3$
2.5 $a^{(2+1+1)}b^2 = a^4b^2$
2.6 $x^{(3+2+1)}y^{(1+1)} = x^6y^2$
2.7 $d^{(1+1+3)}e^{(4+1+2)} = d^5e^7$
2.8 $x^{(2+2+1)}y^{(6+1)} = x^5y^7$

Self-test 3

3.1 a^{-3}
3.2 c^{-2}
3.3 $1 \div b^3 = b^{-3}$
3.4 $1 \div d^4 = d^{-4}$
3.5 $z^{-3.3}$
3.6 $1 \div p^{13.3} = p^{-13.3}$
3.7 $1 \div d^5 = d^{-5}$
3.8 $1 \div (j^4k^5h^3) = j^{-4}k^{-5}h^{-3}$

Self-test 4

4.1 $1/2$
4.2 $1/100$
4.3 $1/(100 \times 100) = 1/10\,000$
4.4 $1/(6 \times 6 \times 6) = 1/216$ or 216^{-1}
4.5 $1/5^5 = 1/3125$ or 3125^{-1}
4.6 $1/13^3 = 1/2197$ or 2197^{-1}

Self-test 5

5.1 $10^{(2+4)} = 10^6$
5.2 $10^{(3+5)} = 10^8$
5.3 $10^{(0+2)} = 10^2$
5.4 $10^{(20+40)} = 10^{60}$
5.5 $6^{(3+12)} = 6^{15}$
5.6 $7^{(2+4)} = 7^6$
5.7 $b^{(9+2)} = b^{11}$
5.8 $z^{(15+2)} = z^{17}$
5.9 $b^{(4.1+7.2+3.8)} = b^{15.1}$
5.10 $k^{(6.22+8.12)} = k^{14.34}$

Self-test 6

6.1 $10^{(2-4)} = 10^{-2}$

6.2 $10^{(3-5)} = 10^{-2}$

6.3 $10^{(0-12)} = 10^{-12}$

6.4 $10^{(-2-4)} = 10^{-6}$

6.5 $4^{(3-12)} = 4^{-9}$

6.6 $1.5^{(2-4)} = 1.5^{-2}$

6.7 $b^{(9-(-2))} = b^{11}$

6.8 $z^{(5-2.5)} = z^{2.5}$

6.9 $z^{(1.3-2.7)} = z^{-1.4}$

6.10 $c^{(3.15-2.93)} = c^{0.22}$

Self-test 7

7.1 $10^{(2 \times 5)} = 10^{10}$

7.2 $10^{(3 \times 10)} = 10^{30}$

7.3 $a^{(7 \times 3)} = a^{21}$

7.4 $a^{(3 \times 7)} = a^{21}$

7.5 $p^{(4.4 \times 1.2)} = p^{5.28}$

7.6 $7^{(3.3 \times 7.8)} = 7^{25.74}$

Self-test 8

8.1 $27^{1/3}$

8.2 $16^{1/2}$

8.3 $p^{1/4}$

8.4 $(7b)^{1/9}$

8.5 $12^{1/j}$

8.6 $k^{1/i}$

Chapter 11:
Powers II: functions of exponentials and logarithms

Self-test 1

1.1 20.09

1.2 24.53

1.3 1.203×10^6

1.4 0.0408

1.5 $K = \exp(-(-12\ 000)/(8.314 \times 298)) = \exp(+(12\ 000/2478)) = \exp(4.843)$
    $= 126.9$

1.6 $K = \exp(-(-40\ 200)/(8.314 \times 298)) = \exp(+(40\ 200/2478)) = \exp(16.223)$
    $= 1.110 \times 10^7$

Self-test 2

2.1 $e^{\ln x} = e^7 \quad \rightarrow \quad x = e^7$

2.2 $e^{\ln x^2} = e^y \quad \rightarrow \quad x^2 = e^y \quad \rightarrow \quad x = \sqrt{e^y}\, \exp(y/2)$

2.3 $e^{\ln xt} = e^y \quad \rightarrow \quad xt = e^y \quad \rightarrow \quad e^y/t$

2.4 $\ln(e^x) = \ln 3 \quad \rightarrow \quad x = \ln 3$

2.5 $\ln(-6x) = \ln h \quad \rightarrow \quad -6x = \ln h \quad \rightarrow \quad \ln h/-6$

2.6 $\ln(e^{(x^2)}) = \ln y \quad \rightarrow \quad x^2 = \ln y \quad \rightarrow \quad \sqrt{\ln y}$

Self-test 3

3.1 4.61

3.2 3

3.3   3.91

3.4   1.531

3.5   Display will say 'error' because it's impossible to conceive of an index so small that e^x can have a value of 0.

3.6   1

3.7   Display will say 'error' because it's impossible for any index to change the overall sign of e^x.

3.8   0

Self-test 4

4.1 $\ln(5 \times 3) = \ln 15$

4.2 $\ln(5 \times 8 \times 2) = \ln 80$

4.3 $\log(20 \times 7) = \log 140$

4.4 $\log_p 7ab$

4.5 $\log ghj$

4.6 $\log nmp$

4.7 $\log_6 qr$

4.8 $\log_n(6 \times 4 \times 2 \times ft) = \log_n(48ft)$

Self-test 5

5.1 $\ln(6/3) = \ln 2$

5.2 $\ln(5f/5) = \ln f$

5.3 $\log(12/4) = \log 3$

5.4 $\log(y^2/y) = \log y$

5.5 $\log(6g/3g) = \log 2$

5.6 $\log(h/p)$

Self-test 6

6.1 $5 \log a = \log a^5$

6.2 $\log(b \times b^2) = \log b^3$

6.3 $\log(c^3 \times c) = \log c^4$

6.4 $\ln(c^2)^3 = \ln c^6$

6.5 $\ln y^6 + \ln(4y)^2 = \ln(y^6 \times 4^2y^2) = \ln 16y^8$

Chapter 12:
Powers III: obtaining linear graphs from non-linear functions

Self-test 1

		$y =$	$m \times$	$x +$	c
1.1	The Nernst equation:	$E_{Cu^{2+}, Cu}$	$\dfrac{RT}{2F}$	$\ln[Cu^{2+}]$	$E^{\ominus}_{Cu^{2+}, Cu}$
1.2	The Clausius–Clapeyron equation:	$\ln p$	$-\dfrac{\Delta H^{\ominus}_{(vap)}}{R}$	$\dfrac{1}{T}$	c
1.3	The second-order rate equation:	$\dfrac{1}{[C]_{t=t}}$	k	t	$\dfrac{1}{[C]_{t=start}}$
1.4	Boyle's law:	p	c	$\dfrac{1}{V}$	(i.e. the constant $= 0$)
1.5	The van't Hoff isochore:	$\ln K$	$-\dfrac{\Delta H^{\ominus}}{R}$	$\dfrac{1}{T}$	c

Self-test 2

The graph below of $\ln \eta$ (as y) against $\ln M$ (as x) is linear. Its gradient is 0.63 and its intercept is -10.2. We see α is 0.63 and $K = \exp(-10.2)$, i.e. $K = 3.8 \times 10^{-5}$. Therefore, for polystyrene in benzene, the **Mark–Houwink equation** has the form $\eta = 3.8 \times 10^{-5} M^{0.63}$.

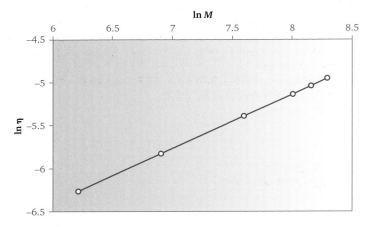

Self-test 3

The graph below of $\ln I$ (as y) against $\ln \omega$ (as x) is linear. Its gradient is 0.50, so the relationship is $I = k\omega^{1/2}$, which we could have written as $I = k\sqrt{\omega}$. (This square-root relationship underpins the **Levich** equation.)

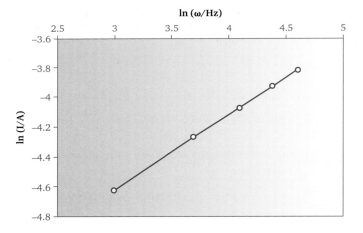

Chapter 13:
Trigonometry

Self-test 1

1.1	0.809
1.2	0.839
1.3	0.213
1.4	0.5
1.5	0.707
1.6	5.67

Self-test 2

2.1 $\theta = \sin^{-1}(0.3) \quad \rightarrow \quad \theta = 17.4°$

2.2 $\theta = \cos^{-1}(0.92) \quad \rightarrow \quad \theta = 23.1°$

2.3 $2\cos\theta = 0.4 \quad \rightarrow \quad \cos\theta = (0.4 \div 2) = 0.2 \quad \rightarrow \quad \theta = \cos^{-1}(0.2), \text{ so } \theta = 78.5°$

2.4 $\dfrac{1}{2}\sin\theta = 0.45 \quad \rightarrow \quad \sin\theta = (2 \times 0.45) = 0.9 \quad \rightarrow \quad \theta = \sin^{-1}(0.9), \text{ so } \theta = 64.2°$

2.5 $\cos\theta + 2 = 2.1 \quad \rightarrow \quad \cos\theta = 0.1 \qquad\qquad \rightarrow \quad \theta = \cos^{-1}(0.1), \text{ so } \theta = 84.2°$

2.6 $\sin^2\theta = 0.9 \quad \rightarrow \quad \sin\theta = \sqrt{0.9} = 0.948 \quad \rightarrow \quad \theta = \sin^{-1}(0.948), \text{ so } \theta = 71.4°$

2.7 $2\tan^2\theta = 9 \quad \rightarrow \quad \tan^2\theta = (9 \div 2) = 4.5 \quad \rightarrow \quad \tan\theta = \sqrt{4.5} = 2.12 \quad \rightarrow$
$\theta = \tan^{-1}(2.12) = 64.7°$

2.8 $4\sin^3\theta = 0.76 \quad \rightarrow \quad \sin^3\theta = (0.76 \div 4) = 0.19 \quad \rightarrow \quad \sin\theta = \sqrt[3]{0.19}$
$= 0.575 \quad \rightarrow \quad \theta = \sin^{-1}(0.575) = 35.1°$

Self-test 3

3.1 $\pi/3$ rad

3.2 $5\pi/6$ rad

3.3 $0.88\,\pi$ rad $= 2.79$ rad

3.4 $0.094 \times 2\pi = 0.593$ rad

Self-test 4

4.1 $c^2 = 2^2 + 5^2 \qquad \rightarrow \quad 4 + 25 = 29 \qquad\qquad \rightarrow \quad c = \sqrt{29} = 5.39$

4.2 $c^2 = 7^2 + 13^2 \qquad \rightarrow \quad 49 + 169 = 218 \qquad\quad \rightarrow \quad c = \sqrt{218} = 14.8$

4.3 $c^2 = 150^2 + 140^2 \quad \rightarrow \quad 22\,500 + 19\,600 = 42\,100 \quad \rightarrow \quad c = \sqrt{42\,100} = 205\,\text{pm}$

Self-test 5

5.1 $c^2 = 4^2 + 7^2 - (2 \times 4 \times 7) \times \cos 73° \rightarrow c^2 = 16 + 49 - 56\cos 73° \rightarrow$
$c^2 = 65 - (56 \times 0.292) \rightarrow c^2 = 65 - 16.4 \rightarrow c^2 = 48.63 \rightarrow c = 6.97$

5.2 $8^2 = 5^2 + 4^2 - (2 \times 5 \times 4\cos\theta) \quad \rightarrow \quad 64 = (25 + 16) - 40\cos\theta \rightarrow \quad 64 = 41 - 40\cos\theta$
$\rightarrow (64 - 41) = -40\cos\theta \quad \rightarrow \dfrac{23}{40} = \cos\theta \rightarrow \quad \theta = \cos^{-1}(-0.578) \rightarrow \quad \theta = 125°$

Chapter 14:
Differentiation I: rates of change, tangents, and differentiation

Self-test 1

1.1 gradient $= 6$

1.2 gradient $= 48$

1.3 gradient $= 5$

1.4 gradient $= 16$

Self-test 2

2.1 $dy/dx = 2x$

2.2 $dy/dx = 3x^2$

2.3 $dy/dx = 12x^2$

2.4 $dy/dx = 3x^2$

Self-test 3

3.1 $dy/dx = 6x^5$

3.2 $dy/dx = 12x^{11}$

3.3 $dy/dx = 5x^4$

3.4 $dy/dx = 3x^2$

3.5 $dy/dx = 1$

3.6 $dy/dx = -2x^{-3}$

3.7 $dy/dx = -11x^{-12} = -11/x^{12}$

3.8 $dy/dx = -7x^{-8} = -7/x^8$

3.9 $dy/dx = -3x^{-4} = -3/x^4$

3.10 $dy/dx = 2.73x^{1.73}$

Self-test 4

4.1 $dy/dx = 9x^2$

4.2 $dy/dx = 70x^{13}$

4.3 $dy/dx = 20x^4$

4.4 $dy/dx = -8.4x^{-8}$

4.5 $dy/dx = 25.8x^5$

4.6 $dy/dx = -(4 \times 10^6)x^{-5}$

Self-test 5

5.1 $dy/dx = 9x^2$

5.2 $dy/dx = 4$

5.3 $dy/dx = 1.6x^{-0.5}$

5.4 $dy/dx = 1.2 \times 10^6 \, x$

5.5 $dy/dx = 1.55x^{-2/3} = 1.55/x^{-2/3}$

5.6 $dy/dx = 92x^{8.2}$

Self-test 6

6.1 $dy/dx = 3x^2$

6.2 $dy/dx = 4x$

6.3 $dy/dx = +0.5x^{-0.5} = -1/2\sqrt{x}$

6.4 $dy/dx = -3x^{-4} = -3/x^4$

6.5 $dy/dx = x$

Self-test 7

7.1 $dy/dx = 4x^3 + 2x$

7.2 $dy/dx = 3x^2 + 4x^{-5}$

7.3 $dy/dx = 7x^6 - 12x$

7.4 $dy/dx = 0.5x^{-0.5} + 12$

Chapter 15:
Differentiation II: differentiating other functions

Self-test 1

1.1 $dy/dx = 5 \, e^{5x}$

1.2 $dy/dx = 3.4 \, e^{3.4x}$

1.3 $dy/dx = 7f \, e^{fx}$

1.4 $dy/dx = 39.06 \, e^{4.2x}$

1.5 $dy/dx = d^2 \, e^{dx}$; the factor of d^2 comes from $d \times d \, e^{dx}$.

1.6 $dy/dx = (3.492 \times 10^{-6}) \, e^{(4\times10^{-7})x}$

1.7 $dy/dx = e^x + e^{-x}$

1.8 $dy/dx = 12x^3 - 14x + 5 \, e^{5x}$

Self-test 2

2.1 $dy/dx = 1/x$
2.2 $dy/dx = 1/x$
2.3 $dy/dx = 5.4/x$
2.4 $dy/dx = -1/x$
2.5 $dy/dx = 1/x$
2.6 $dy/dx = ab/x$
2.7 $dy/dx = 4 - 3/x$
2.8 $dy/dx = 4x^3 + 3/x^4 + 1/x$
2.9 $dy/dx = 1000 + 1/x$
2.10 $dy/dx = 1/x - c/x$

Self-test 3

3.1 $dy/dx = 4\cos 4x$
3.2 $dy/dx = 12\ \cos\ 3x$
3.3 $dy/dx = -97.2\cos(8.1\ x)$
3.4 $dy/dx = -44\ \sin(44x)$
3.5 $dy/dx = 49.14\ \sin(-7.8\ x)$
3.6 $dy/dx = -d^2 \cos(dx)$; the factor of d^2 here comes from $d \times d\ \cos(dx)$.
3.7 $dy/dx = \cos\ x - \sin\ x$
3.8 By saying $\cos x/2 = \dfrac{1}{2}\cos\ x$ and $(\sin 3x)/4 = \dfrac{1}{4}\ \sin 3x$, so $dy/dx = -\dfrac{1}{2}(\sin x)$ $-\ 3 \times \dfrac{1}{4}(3\cos 3x)$.

Self-test 4

4.1 $dy/dx = -4 \times 5\ \sin\ 5x$, so gradient $= 20\ \sin(5\pi/6) = 20 \times -0.5 = -10$
4.2 $dy/dx = 2 \times 7\ \cos\ 7x$, so gradient $= 14 \sin(7\pi/8) = 5.356$

Chapter 16:
Differentiation III: differentiating functions of functions: the chain rule

Self-test 1

1.1 $(dy/dx) = 4x(x^2 + 2)$
1.2 $(dy/dx) = \dfrac{2}{x}\ln x$
1.3 $(dy/dx) = 3\,e^{3x} \times\ \cos(e^{3x})$
1.4 $(dy/dx) = 4x^3 e(x^4)$
1.5 $(dy/dx) = 3x^2/x^3 = 3/x$
1.6 $(dy/dx) = 5\ \cos x(\sin\ x)^4 \equiv 5\ \cos x\ \sin^4 x$
1.7 $(dy/dx) = -(4x^3 - 2x)\sin(x^4 - x^2)$
1.8 Remember that $\sqrt{x} = x^{0.5}$ and $1/\sqrt{x} = x^{-0.5}$:
$$\left(\frac{dy}{dx}\right) = \frac{3}{2} \times \frac{x^{-0.5}}{x^{0.5}}$$
The fraction: $x^{-0.5}/x^{0.5} \equiv 1/\sqrt{x} \times 1/\sqrt{x} = 1/x$

so the differential becomes $3/2 \times 1/x$, i.e. $3/2x$.

1.9 $(dy/dx) = 9/2x^{0.5} \times 1/3x^{3/2} = (3/2)(x^{0.5}/x^{3/2}) = 3/2x$
1.10 $(dy/dx) = (-6/x^3 + 6/x^4) \times \exp(3/x^2 - 2/x^3)$

Chapter 17:
Differentiation IV: the product rule and the quotient rule

Self-test 1

1.1   We start by writing the logarithm function within brackets, in order to separate it from the power term: $y = x^4(\ln x)$. There is no further means of simplification, so we see how there are two separate functions, so y here is a product.

1.2   Using the second law of powers (Chapter 10), the two powers can cancel: $y = x^{(3-5)} = x^{-2}$, so y is a simple function.

1.3   It is not possible to cancel this expression or to simplify it further. Accordingly, it is a quotient.

1.4   Using the first law of powers (see Chapter 10), we can rewrite the expression as $y = 4 \times x^{(4+5)} = 4x^9$. Accordingly, y is a simple, single function.

1.5   The two functions $\sin x$ and $\cos x$ are separate, so y is a product.

1.6   We start by writing the sine function in brackets to separate it from the power term, so $y = x^3(\sin 3x^2)$. There is no further means of simplification, so we see how there are two separate functions, so y here is a product.

1.7   We start by recognizing two functions, both sin and ln. There is no further means of simplification, so we see how there are two separate functions, so y here is a product.

1.8   It is not possible to cancel this expression or to simplify it further. Accordingly, it is a quotient.

Self-test 2

	u	$\dfrac{du}{dx}$	v	$\dfrac{dv}{dx}$	$\dfrac{dy}{dx}$ obtained via the product rule (The first expression employs square brackets [ ] to indicate the two differential terms. The subsequent expressions use curly brackets { } to indicate factorizing.)
2.1	$2x^4$	$8x^3$	$\sin 3x$	$3 \cos 3x$	$2x^4 \times [3 \cos 3x] + \sin 3x \times [8x^3]$ $= 2x^3\{3x \cos 3x + 4\sin 3x\}$
2.2	$\ln 2x$	$1/x$	$\exp(2x)$	$2 \exp(2x)$	$\ln 2x \times [2\exp(2x)] + \exp(2x) \times [1/x]$ $= \exp(2x)\{2 \ln 2x + (1/x)\}$
2.3	x^5	$5x^4$	$\ln x$	$1/x$	$x^5 \times [1/x] + \ln x \times [5x^4] = x^4\{1 + 5 \ln x\}$
2.4	$2x^{-4}$	$-8x^{-5}$	$\cos 4x$	$-4 \sin 4x$	$2x^{-4} \times [-4 \sin 4x] + \cos 4x \times [-8x^{-5}]$ $= -8x^{-5}\{x \sin 4x + \cos 4x\}$
2.5	$\sin^2 x$	via the chain rule: $2 \sin x \times \cos x$	$\cos^2 x$	via the chain rule: $-2 \sin x \times \cos x$	$\sin^2 x \times [-2 \sin x \cos x] + \cos^2 x$ $\times [2 \sin x \cos x] = 2 \sin x \cos x\{\cos^2 x - \sin^2 x\}$
2.6	$\ln x^2$	via the chain rule: $1/x^2 \times 2x = 2/x$	$\sin^3 2x$	via the chain rule: $3 \times 2 \sin^2 2x$ $\cos 2x$	$\ln x^2 \times [6 \sin 2x \cos 2x] + \sin^3 2x \times [2/x]$ $= 2 \sin^2 2x\{3 \ln x^2 \cos 2x + (\sin 2x/x)\}$

Self-test 3

	u	$\dfrac{du}{dx}$	v	$\dfrac{dv}{dx}$	$\dfrac{dy}{dx}$ obtained via the product rule (The first expression employs square brackets [ ] to indicate the two differential terms. The subsequent expressions use curly brackets { } to indicate factorizing.)
3.1	x^2	$2x$	$\sin x$	$\cos x$	$\dfrac{\sin x[2x] - x^2[\cos x]}{(\sin x)^2} = \dfrac{x\{2\sin x - x\cos x\}}{\sin^2 x}$
3.2	$\ln x$	$1/x$	x^4	$4x^3$	$\dfrac{x^4[1/x] - \ln x[4x^3]}{(x^4)^2} = \dfrac{x^3 - 4x^3\ln x}{x^8} = \dfrac{x^3\{1 - 4\ln x\}}{x^8}$ $= \dfrac{1 - 4\ln x}{x^5}$
3.3	$\sin x$	$\cos x$	$\cos x$	$-\sin x$	$\dfrac{\cos x[\cos x] - \sin x[-\sin x]}{\sin^2 x} = \dfrac{\cos^2 x + \sin^2 x}{\cos^2 x}$
3.4	$\exp(2x)$	$2\exp(2x)$	$3x^3$	$9x^2$	$\dfrac{3x^3[2\exp(2x)] - \exp 2x[9x^2]}{(3x^3)^2} = \dfrac{3x^2\exp 2x\{2x - 3\}}{9x^6}$ $= \dfrac{\exp 2x\{2x - 3\}}{3x^4}$

Chapter 18:
Differentiation V: maxima and minima on graphs: second differentials

Self-test 1

1.1 The function x^4 is a simple, single-term polynomial function, and will therefore have one turning point.

1.2 The equation is a simple quadratic, and will therefore have a single turning point.

1.3 This long expression involves a series of polynomials. The highest power is 4, i.e. x^4, so we expect up to $(4 - 1) = 3$ turning points.

1.4 Another long expression involving a series of polynomials. The highest power is 5, i.e. x^5, so we expect up to $(5 - 1) = 4$ turning points.

Self-test 2

2.1 $dy/dx = 2x + 1$. At the turning point, $0 = 2x + 1$, so $x = -\dfrac{1}{2}$. Back-substitution yields $y = \left(-\dfrac{1}{2}\right)^2 + \left(-\dfrac{1}{2}\right) + 5$, so $y = 4.75$.

2.2 $dy/dx = 6x - 12$. At the turning point, $0 = 6x - 12$, so $x = 2$. Back-substitution yields $y = (2)^2 + 2 - 7$, so $y = -1$.

2.3 $dy/dx = -10x + 4$. At the turning point, $0 = -10x + 4$, so $x = 0.4$. Back-substitution yields $y = -5 \times (0.4)^2 + 4 \times (0.4) + 2$, so $y = 4.4$

2.4 $dy/dx = 6.4x - 9.1$. At the turning point, $0 = 6.4x - 9.1$, so $x = 1.421$. Back-substitution yields $y = -3.2 \times (1.421)^2 - 9.1 \times (1.421) = 6.462 - 12.931$, so $y = -6.469$.

2.5 $dy/dx = x^2 + 10x + 24$, which factorizes to $dy/dx = (x + 6)(x + 4)$. At the turning point, therefore, $0 = (x + 6)(x + 4)$, and $x = -6$ or -4. Back-substitution allows us to compute the turning points as: $(-6, -29)$ and $(-4, -30.3)$.

2.6 $dy/dx = 3x^2 + 18x + 24$, which factorizes to $dy/dx = 3(x + 2)(x + 4)$. At the turning point, therefore, $0 = (x + 2)(x + 4)$. Back-substitution yields the turning points as: $(-2, -20)$ and $(-4, -16)$.

Self-test 3

4.1 $dy/dx = 3x^2$, turning point $= (0,2)$ $d^2y/dx^2 = 6x$, i.e. 0 at $x=0$, so an inflection

4.2 $dy/dx = 6x + 6$, turning point $= (-1,-3)$ $d^2y/dx^2 = 6$, i.e. positive, so a minimum

4.3 $dy/dx = 12x^2 + 8x = 4x(3x + 2)$ and $d^2y/dx^2 = 24x + 8$

Turning points occur either when $4x=0$, i.e. when $x=0$, or when $(3x+2)=0$, i.e. when $x=-2/3$.

First turning point $= (0, 12)$, so $d^2y/dx^2 = 8$ when $x=0$, i.e. positive, so a minimum.

Second turning point $= (-2/3, 11.66)$, so $d^2y/dx^2 = -8$ when $x = -2/3$, i.e. negative, so a maximum.

Chapter 19:
Integration I: reversing the process of differentiation

Self-test 1

1.1 $y = \dfrac{1}{4} x^4 = x^4/4$

1.2 $y = 2x^6/6 = x^6/3$

1.3 $y = x^{2.3}/2.3$

1.4 $y = (3.7\, x^{9.2})/9.2$

1.5 $y = \dfrac{6}{-2x^2} = -(3/x^2)$

1.6 $1/x^{4.2} = x^{-4.2}$, so $y = -1/(3.2\, x^{3.2}) + 5x$

1.7 $y = (x^3/3) + (3/2)x^2 - 6x$

1.8 $4/x^5 = 4x^{-5}$ and $3.5/x^2 = 3.5\, x^{-2}$, so $y = (-4/4x^4) - (-3.5/x) = -(1/x^4) + (3.5/x)$

1.9 $\sqrt{x^7} = x^{7/2}$ so $y = (x^{9/2}/(9/2)) = (2\, x^{9/2})/9$

1.10 $\sqrt{x^6} = x^{6/2} = x^3$ so $y = x^4/4 - 2\, x^3/3$

Self-test 2

2.1 $y = \dfrac{1}{4} \sin 4x$

2.2 $y = -6/4 \cos 4x$

2.3 $y = -5/2x^{-2}$

2.4 $y = 3 \ln x$

2.5 $y = e^{5x}/10$

2.6 $y = 96.4x$

2.7 $y = x^5/5 + y = \dfrac{1}{2} \cos 2x$

2.8 $y = e^{dx}/d + e^{ex}/e$

2.9 $y = 5 \ln x + e^{3.2x}/3.2$

2.10 $y = b/a \cos ax + a/b \sin bx = \dfrac{b^2 \cos ax + a^2 \sin bx}{ab}$

Self-test 3

3.1 The equation of the line is $x^4/4 + c$. Substituting yields $c = 9$.

3.2 The equation of the line is $e^{3x}/3 + c$. Substituting yields $c = -18.1$.

3.3 The equation of the line is $x^{1.5}/1.5 + c$. Substituting yields $c = -9.3$.

3.4 The equation of the line is $y = \ln x + c$. Substituting yields $c = 3.61$.

3.5 The equation of the line is $y = (x^4/4) - (x^3/3) + c$. Substituting yields $c = 0.75$.

3.6 The equation of the line is $y = 3 \ln x + c$. Substituting yields $c = 0.92$.

Chapter 20:
Integration II: separating the variables, integration with limits, and area determination

Self-test 1

1.1 $dy = zx^3 \, dx \rightarrow \int dy = \int zx^3 \, dx \rightarrow y = \frac{zx^4}{4} + \text{constant}$

1.2 $\frac{1}{y^2}\frac{dy}{dx} = x^2 \rightarrow \int \frac{1}{y^2} \, dy = \int x^2 \, dx \rightarrow -\frac{1}{y} = \frac{x^3}{3} + \text{constant}$

1.3 $\frac{1}{c^3}\frac{dc}{de} = \frac{1}{e} \rightarrow \frac{1}{c^3} \, dc = \frac{1}{e} \, de \rightarrow \int \frac{1}{c^3} \, dc = \int \frac{1}{e} \, de \rightarrow -\frac{1}{2c^2} = \ln e + \text{constant}$

1.4 $\frac{1}{h^3}\frac{dh}{dg} = 4\sin g \rightarrow \frac{1}{h^3} \, dh = 4\sin g \, dg \rightarrow \int \frac{1}{h^3} \, dh = \int 4\sin g \, dg \rightarrow$

$-\frac{1}{2h^2} = -4\cos g + \text{constant} \rightarrow \frac{1}{h^2} = 8\cos g + \text{constant}$

Self-test 2

2.1 $y = \left[\frac{x^5}{5} - \frac{x^3}{3}\right]_1^2 = \left[\frac{32}{5} - \frac{8}{3}\right] - \left[\frac{1}{5} - \frac{1}{3}\right] = (6.4 - 2.7) - (0.2 - 0.33) = 3.86$

2.2 $y = \left[\frac{e^{3x}}{3}\right]_1^3 = \left[\frac{e^9}{3}\right] - \left[\frac{e^3}{3}\right] = \frac{8103}{3} - \frac{20}{3} = 2701 - 7 = 2694 \text{ (to zero d.p.)}$

2.3 $y = \left[\frac{\sin 2x}{2}\right]_{\pi/3}^{\pi} = \left[\frac{\sin 2\pi}{2}\right] - \left[\frac{\sin 2\pi/3}{2}\right] = 0 - 0.433 = -0.433$

2.4 $y = \left[\frac{x^4}{4} - \frac{x^3}{3} + \frac{x^2}{2} - 6x\right]_0^5 = \left[\frac{5^4}{4} - \frac{5^3}{3} + \frac{5^2}{2} - 30\right] - 0 = 625.8$

2.5 $y = [2 \ln x]_{9.5}^{10} = 2 \ln\left(\frac{10}{9.5}\right) = 2 \ln(1.053) = 2 \times 0.052 = 0.103$

2.6 $y = \left[\frac{\sin 6x}{18}\right]_0^{\pi} = \left[\frac{\sin 6\pi}{18}\right] - \left[\frac{\sin 0}{18}\right] = 3 \times 2\pi \text{ rad so } \sin 6\pi = 0. \quad \sin\theta = 0, \text{ so } \int = 0$

Self-test 3

3.1 $\text{area} = \left[\frac{x^3}{3} + 4x\right]_3^5 = \left[\frac{125}{3} + 20\right] - \left[\frac{27}{3} - 12\right] = [41\frac{2}{3} + 20] - [9 + 12] = 64\frac{2}{3}$

3.2 $\text{area} = -\left[\frac{1}{x}\right]_2^6 = -\left[\frac{1}{6} - \frac{1}{2}\right] = \frac{1}{3}$

Chapter 21:
Statistics I: averages and simple data analysis

Self-test 1

1.1 $\bar{m} = 11.97\,g$
1.2 $\bar{c} = 691.8\,ppm$
1.3 $\overline{\Delta H} = 32.16\,kJ\,mol^{-1}$

Self-test 2

2.1 The mass 9.21 mg occurs three times, so it is the mode.
2.2 The concentration 121 ppm occurs three times, so it is the mode.
2.3 The concentration 33 μg dm^{-3} occurs four times, so it is the mode.

Self-test 3

3.1 The data set comprises nine data items, so the median value is the fifth. When placed in order, the median is therefore 1.93 cm.

3.2 The data set comprises six data items, so the median is the mean of items 3 and 4. The median is therefore the mean of 474 and 475, i.e. 474.5 kJ mol^{-1}.

Self-test 4

$$s = 0.033 \, \text{cm} \qquad 4.2 \quad s = 1.47 \, \text{kJ mol}^{-1}$$

(Each data set comprises fewer than 10 entries, so the correct standard deviation is the *sample* standard deviation, s.)

Self-test 5

	N	x	$x_{(next)}$	$x_{(minimum)}$	$x_{(maximum)}$	$Q_{(exp)}$	Confidence to reject?
5.1	5	**0.92**	0.72	0.65	0.92	0.714	$95 < \% < 99$
5.2	4	**226**	219	214	226	0.583	less than 90%
5.3	6	**32.2**	35.4	32.2	35.7	0.914	greater than 99%

Chapter 22:
Statistics II: treatment and assessment of errors

Self-test 1

1.1 2.4:1 (not safe)
1.2 27.8:1 (safe)
1.3 125:1 (safe)
1.4 29.7:1 (safe)
1.5 92:1 (safe, because 12 mV = 0.012 V)
1.6 68:1 (safe)

Self-test 2

2.1 Using eqn (22.3), the minimum error = (1 mm ÷ 112 mm) = 8.93×10^{-3}, or 0.9%.

2.2 Using eqn (22.3), the minimum error = (0.5°C ÷ 24.9 °C) = 0.020, or 2.0%.

2.3 Using eqn (22.4), $(\text{minimum error})^2 = \left(\dfrac{0.1°\text{C}}{0.7°\text{C}}\right)^2 + \left(\dfrac{0.001 \, \text{g}}{3.503 \, \text{g}}\right)^2$

$(\text{minimum error})^2 = 0.02041 + 2.855 \times 10^{-4}$

$(\text{minimum error})^2 = 0.02068$
so: minimum error $= \sqrt{0.02041}, = 0.1428$

The minimum error represents about 14%. The overwhelming majority of this error comes from the reading of temperature. The chemist should use a superior thermometer, such as a Beckmann thermometer, or a sensitive thermocouple.

2.4 Using eqn (22.4):

$$(\text{minimum error})^2 = \left(\frac{0.001 \, \text{g}}{0.504 \, \text{g}}\right)^2 + \left(\frac{0.5 \, \text{cm}^3}{100 \, \text{cm}^3}\right)^2 + \left(\frac{0.5 \, \text{cm}}{10 \, \text{cm}}\right)^2$$
$$+ \left(\frac{0.05}{0.74}\right)^2$$

$$(\text{minimum error})^2 = 3.9368 \times 10^{-6} + 2.500 \times 10^{-5} + 2.500 \times 10^{-3} + 4.565 \times 10^{-3}$$

$$(\text{minimum error})^2 = 7.0939 \times 10^{-3}$$

so: minimum error $= \sqrt{7.0939 \times 10^{-3}} = 0.0842$

The minimum error is 8.4%. The principal error is the absorbance reading. The chemist should use a superior instrument; for example, the innate error in the absorbance reading with a good, modern instrument can be as low as 0.001.

Self-test 3

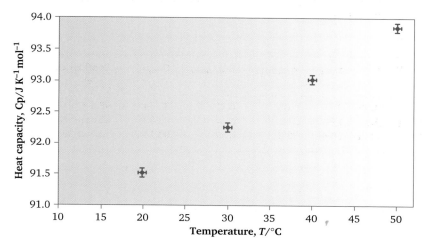

Self-test 4

The graph below shows some slight scatter, and with a correlation coefficient r of -0.9910. The negative sign is a direct consequence of the negative slope. The value of r^2 is therefore $(-0.9910)^2 = 0.982$.

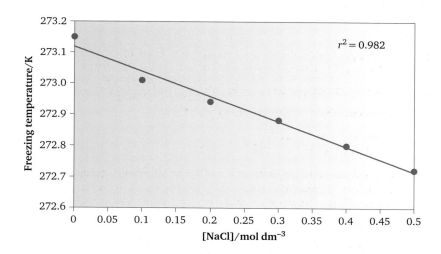

Bibliography

Books of mathematics for chemists

Chemical Calculations, Paul Yates, Chapman and Hall, London, 1997. This is Yates' first maths book. It is relatively short, and aimed at those with little or no maths. Yates' style is gentle and persuasive. It *encourages* the reader. He also employs a margin icon of a calculator, demonstrating its correct use at appropriate junctures.

Yates is a British physical chemist; the chapters of his book relate to a traditional course in physical chemistry (level I and some level II) so, within each chapter, Yates discusses the mathematics needed by a typical student. For example, Chapter 3 is entitled 'Solution Chemistry', Chapter 4 is 'Kinetics', and so on. This intelligent approach works well for students who struggle during a course in physical chemistry, but may not work so well for students on courses explicitly aimed at teaching mathematics. Yates himself must have felt the approach failed somewhat, because he includes an appendix of pure mathematics. Also, there is no logical progression from the mathematical fundamentals via symbols through to algebra, powers, etc.

Chemical Calculations at a Glance, Paul Yates, Blackwell, Oxford, 2005. This book represents a 're-statement' of Yates' earlier book *Chemical Calculations* (as above) since it covers the same material but is themed mathematically rather than according to the syllabus of a traditional course in physical chemistry.

The book is beautifully produced in A4 format, and shows a logical layout throughout, starting with notation and units, and moving eventually to calculus. The proportion of chemical examples is high. The number of worked examples is similarly higher than the average, although occasionally more detail would have helped. Nevertheless, it's a helpful tool.

The Chemistry Maths Book, Erich Steiner, Oxford University Press, Oxford, 1996. This book represents the 'Rolls-Royce' end of the market. Steiner is an inspiring and gifted teacher, and this book reflects that gift. It is beautifully written and well produced. The text is authoritative, comprehensive. It also has a good, long index. Furthermore, Steiner litters the text with clever quotes; most pages glisten with fascinating historical detail in a series of well-designed footnotes.

Maths for Chemists, Volume 1: Numbers, Functions and Calculus, Martin C. R. Cockett and Graham Doggett, RSC, Cambridge, 2003. This is a good book. It is well written and produced, as are all titles in the RSC's new 'Tutorial Chemistry Texts' series. The chapter on algebra is poor and far too short at 10 pages. There are no chemical structures, although to be fair the proportion of chemistry is pleasingly high. The index is short but better than average.

Maths for Chemists, Volume 2: Power Series, Complex Numbers and Linear Algebra, Martin C. R. Cockett and Graham Doggett, RSC, Cambridge, 2003. This book is the companion volume to the text immediately above. By 'linear algebra', the authors mean vectors, arrays, determinants, and the like. The book is very advanced, though its treatment of quite difficult topics is superb.

Beginning Mathematics for Chemists, Stephen K. Scott, Oxford University Press, Oxford, 1995. Chemistry staff like this book. Its approach is logical and gentle on the student. I like its principal titles 'warming up' and 'relaxing down', for example. The book is rather too short. Calculus occupies centre stage, with about 50 pages. Algebra occupies a mere 21 pages.

Essential Mathematics for Chemists, John Gormally, Prentice Hall, Harlow, 2000. The book is printed with two colours, and is visually appealing. Gormally arranges the seven short chapters into sensible and predictable groups, starting with 'handling numbers', then 'handling algebra'. It has separate chapters to describe differential and integral calculus. Its index is poor, but does include chemical terms as well as mathematics. The exercises are generally chemical, although almost all of these chemical terms relate only to physical chemistry.

Unfortunately, once more this book looks like a vehicle for mathematicians rather than for chemists. To the eye, its pages look to be crammed with equations. The number of figures is minimal, and there are no chemical structures.

Basic Mathematics for Chemists, Peter Tebbutt, Wiley, Chichester, 1994. The text is very traditional in style and content, with monochrome pages and narrow margins. It is slightly long. The overwhelming majority of the book represents mathematical content rather than chemistry.

Tebbutt's book contains more figures than those above, the majority of which are clearly of chemical origin, but, yet again, there are no chemical structures. The very short index includes chemical terms as well as mathematical terms. Most of these chemical terms relate exclusively to physical chemistry.

Calculations for A-Level Chemistry (3rd edn), E. N. Ramsden, Stanley Thornes, Cheltenham, 1994. This book was in print until quite recently, and some shops may still have copies. A revamped edition is expected soon.

The book contains a few chemical structures, and *all* examples derive from chemistry—and from all branches of chemistry, not just physical chemistry. No student could complain that this book was written for mathematicians.

The style is clear, concise, and constructive. Like Yates, Ramsden chooses to teach by chemical discipline rather than mathematical topic, which will preclude its use from most courses of 'Mathematics for chemists'. The intended audiences are students reading for pre-university qualifications, so its standard is only slightly above that of a modern HND.

Books describing the mathematics

The 'teach yourself' series of books are generally excellent. For example, each of the following is superb and highly recommended:

- *Teach Yourself Mathematics*, Trevor Johnson and Hugh Neill, Hodder and Stoughton, London, 2003.
- *Teach Yourself Algebra*, Paul Abbott and revised by Hugh Neill, Hodder and Stoughton, London, 2003.
- *Teach Yourself Trigonometry*, Paul Abbott and revised by Hugh Neill, Hodder and Stoughton, London, 2003.
- *Teach Yourself Calculus*, Paul Abbott and revised by Hugh Neill, Hodder and Stoughton, London, 2003.

Mathematics support

There are several good websites that support a course of 'Mathematics for chemists'. One of the best is the page of 'Mathematics Support Materials' at the University of Plymouth:

http://www.tech.plym.ac.uk/maths/resources/PDFLaTeX/mathaid.html (accessed 12 May 2005), which hosts pdf files of high quality.

Books describing physical chemistry

Most of the chemistry in these pages comes from the physical branch. Some may see this weighting as unfortunate, but Physical chemistry is generally mathematical.

The physical chemistry in these pages is adequately described in straightforward books on physical chemistry. For example, the best-selling textbook of physical chemistry in the world is undoubtedly Atkins' *Physical Chemistry*. The latest edition is the seventh, by P. W. Atkins and Julio de Paula, Oxford University Press, Oxford, 2002. Many students will find it rather mathematical, and its treatment is certainly highbrow. Its 'little brother' is *Elements of Physical Chemistry* (4th edn), P. W. Atkins, Oxford University Press, Oxford, 2005, and is intended to overcome these perceived difficulties by limiting the scope and level of its parent text. Both are thorough and authoritative.

Several texts approach the topic by means of worked examples. *Physical Chemistry*

(2nd edn), C. R. Metz, McGraw-Hill, New York, 1989, is a member of the 'Schaum Out-line Series' of texts, and *Physical Chemistry*, H. E. Avery and D. J. Shaw, Macmillan, Basingstoke, 1989, is part of the 'College Work-out Series'. Both books are crammed with worked examples, self-assessment questions, and hints on how to approach typical questions. Avery and Shaw is one of the few general textbooks on physical chemistry that a non-mathematician can read with ease.

Monk has tried to get away from a traditional approach in his *Physical Chemistry: Understanding our Chemical World*, Wiley, Chichester, 2005. In this book, the examples take centre stage, and the theory is deduced from these examples. The mathematics is explained in minute detail, but without any loss of rigour.

Books describing analytical chemistry

The following contain sufficient material to explain the analytical chemistry described in these pages: *Statistics and Chemometrics for Analytical Che*mistry (5th edn), J. N. Miller and J. C. Miller, Pearson, Harlow, 2005, is well written and offers a well-balanced approach. While the book *is* mathematical, the maths is not overbearing. Alternatively, *Analytical Chemistry* (5th edn.), G. F. Christian, Wiley, New York, 1994, is fairly comprehensive and will cover some of the material covered here. *The Elements of Analytical Chemistry*, S. P. J. Higson, Oxford University Press, Oxford, 2001, is part of the 'Oxford Primer' series, so will be cheap and affordable.

Books describing electrochemistry

Many of the examples come from electrochemistry, which is a notoriously mathematical discipline. Without doubt, the best 'all round' book for the electrochemist, and already regarded as a modern classic, is *Electrochemical Methods* (2nd edn), A. J. Bard and L. R. Faulkner, Wiley, New York, 2005. Its treatment is generally very mathematical.

Alternatively, *Fundamentals of Electroanalysis*, P. M. S. Monk, Wiley, Chichester, 2001, is written as part of a distance-learning package, 'Analytical Texts in the Sciences'. It is therefore brimming with worked examples and illustrations.

Examples cited in the text

Chapter 2

Worked example 2.10:
For full details of the reaction, see: J. Cordaro, J. McClusker, and R. Bergman, Synthesis of mono-substituted 2,2′-bipyridines, *J. Chem. Soc., Chem. Commun.*, 2002, 1496.

Chapter 6

Worked example 6.11:
The fascinating topic of the relationship between the Golden Ratio and τ is described in the web page, *www.austms.org.au/Modules/Fib/fib.pss* (accessed 10 June 2004). Alternatively, see:

http:www.maven.smith.edu/~phyllo/About/fibogolden.html
(accessed 8 July 2005).

Chapter 13

Worked example 13.4:
The adsorption of methyl viologen on platinum is described in the paper by K. Kobayashi, F. Fujisaki, T. Yoshimina, and K. Nik, An analysis of the voltammetric adsorption waves of methyl viologen, *Bull. Chem. Soc. Jpn*, 1986, **59**, 3715.

Worked example 13.5:
An accessible introduction to magic-angle NMR may be found in C. N. Banwell and E. M. McCash, *Fundamentals of Molecular Spectroscopy* (4th edn), McGraw–Hill, Maidenhead, 1994, pp. 274–6.

Chapter 14

Additional problem 14.3:
The data for this example come from *Physical Chemistry: A Molecular Approach*, Donald A. MacQuarrie and John D. Smith, University Science Books, Sausalito, CA, 1997, p. 954.

Additional problem 14.5:
The temperature dependence cited relates to the electron mobility due to ionized-impurity scattering; see *The Physic and Chemistry of Solids*, Stephen Elliott, Wiley, Chichester, 1998, p. 510.

Additional problem 14.5:
The effective potential is described in Atkins, *Physical Chemistry*, p. 369.

Additional problem 14.7:
The expression is given in Atkins, *Physical Chemistry*, p. 738. Scattering in general is discussed from p. 736.

Additional problem 14.8:

Virial coefficients and other alternatives to the ideal-gas equation are discussed in Atkins, *Physical Chemistry*, p. 17.

Additional problem 14.9:

These aspects of molecular interactions may be found in Atkins, *Physical Chemistry*, p. 699.

Additional problem 14.10:

The form given to b here may be found in Atkins, *Physical Chemistry*, p. 842. (Atkins symbolizes it as B.)

Chapter 15

Additional problem 15.9:

The equation for ϕ comes from *Physical Chemistry: A Molecular Approach*, Donald A. MacQuarrie and John D. Smith, University Science Books, Sausalito, CA, 1997, p. 413. In this text, it is symbolized as ϕ_{1s}^{STO}.

Chapter 16

Additional problem 16.2:

The expression is given in Atkins, *Physical Chemistry*, p. 738.

Additional problem 16.3:

Lasers are discussed in Atkins, *Physical Chemistry*, p. 553ff. Non-linear effects and frequency doubling are discussed on p. 559.

Additional problem 16.4:

An accessible introduction to magic-angle NMR may be found in C. N. Banwell and E. M. McCash, *Fundamentals of Molecular Spectroscopy* (4th edn), McGraw–Hill, Maidenhead, 1994, pp. 274–6.

Additional problem 16.7:

The equation for ϕ comes from *Physical Chemistry: A Molecular Approach*, Donald A. MacQuarrie and John D. Smith, University Science Books, Sausalito, CA, 1997, p. 413. In this text, it is symbolized as ϕ_{1s}^{GF}.

Additional problem 16.8:

The example is amended slightly, from *Physical Chemistry* (2nd edn), C. Metz, Schaum Outline Series, McGraw-Hill, New York, 1989, p. 296.

Additional problem 16.9:

The expression linking λ and θ is given in *Physical Chemistry: A Molecular Approach*, Donald A. MacQuarrie and John D. Smith, p. 540.

Additional problem 16.10:

The expression linking λ and θ is explored in *Physical Chemistry: A Molecular Approach*, Donald A. MacQuarrie and John D. Smith, 1197.

Chapter 17

Worked Example 17.4:

The equation comes from C. N. Banwell, *Fundamentals of Molecular Spectroscopy* (3rd edn.), McGraw-Hill, Maidenhead, 1983, p. 157.

Additional problem 17.4:

Shielding and screening are discussed in Atkins, *Physical Chemistry*, p. 261.

Additional problems 17.6, 17.7:

The derivation of these expressions is available in *Physical Chemistry: A Molecular Approach*, Donald A. MacQuarrie and John D. Smith, University Science Books, Sausalito, CA, 1997, pp. 972, 973.

Additional problem 17.8:

The coiling of polymer chains is discussed in Atkins, *Physical Chemistry*, p. 721.

Additional problem 17.10:

The Eyring equation is discussed in Monk, *Physical Chemistry: Understanding Our Chemical World*, pp. 416–20. The Eyring equation is poorly treated in many books of physical chemistry, and may appear differently than here.

Chapter 19

Probably the first person to develop the mathematics of integration was English mathematician, physicist, and philosopher Sir Isaac Newton (1642–1727). However, our current notation (dy/dx for a derivative and the so-called **script-S** for an integral) is that of Gottfried Leibniz (1646–1716). The 'script S' was used merely as an abbreviation for 'sum'. Its calligraphic style followed contemporary handwriting. For more information on Newton and Liebniz, see the website:

  http://www.bbc.co.uk/history/
historic_figures/newton_isaac shtml

  For more information on writing styles, see the interesting detail at:

  http://www.waldenfont.com/public/
dhmanual.pdf
(both sites accessed 31 March 2004).

Additional problem 19.2:

This temperature relationship between C_p and T is discussed in Atkins, *Physical Chemistry*, pp. 50 and 105.

Additional problem 19.3:

The temperature dependence between C_p and T at very low temperatures is mentioned in Atkins, *Physical Chemistry*, p. 51.

Additional problem 19.4:

The temperature dependence of this phase boundary is given in *Physical Chemistry: A Molecular Approach*, Donald A. MacQuarrie and John D. Smith, University Science Books, Sausalito, CA, 1997, p. 951.

Additional problem 19.5:

For more information, including a full derivation of A, see *Physical Chemistry* (3rd edn), Gilbert Castellan, Addison-Wesley, Reading, MA, 1983, p. 671.

Chapter 20

Additional problem 20.1:

This derivation of an expression for ΔG from the appropriate Maxwell relation is discussed in Atkins, *Physical Chemistry*, 128ff., and Monk, *Physical Chemistry: Understanding our Chemical World*, p. 154ff.

Chapter 21

The term 'standard deviation σ' causes much confusion, in part because some workers use different terms to mean σ, and the same terms to mean different statistical parameters. The web page:

http://mathworld.wolfram.com/ StandardDeviation.html

has a concise but useful summary of the different usages (page accessed 28 April 2005).

Chemical index

Chemistry index

Mathematics index